MW00783504

The Hawkins Ranch in Texas

Number 121: Centennial Series of the Association
of Former Students, Texas A&M University

The *Hawkins Ranch* in Texas

From Plantation Times *to the* Present

Margaret Lewis Furse

Margaret Lewis Furse

Texas A&M University Press
College Station

First edition

This paper meets the requirements of ANSI/NISO Z39.48-1992 (Permanence of Paper).
Binding materials have been chosen for durability.

Library of Congress Cataloging-in-Publication Data

Furse, Margaret Lewis, author.
The Hawkins Ranch in Texas : from plantation times to the present /
Margaret Lewis Furse. — First edition.
pages cm — (Centennial series of the Association of Former Students,
Texas A&M University ; number 121)
Includes bibliographical references and index.
ISBN 978-1-62349-110-9 (cloth : alk. paper) — ISBN 978-1-62349-173-4 (e-book)
1. Hawkins Ranch (Tex.)—History. 2. Plantation life—Texas—Matagorda County. 3. Ranch life—Texas—Matagorda County. 4. Ranchers—Texas—Matagorda County—Biography. 5. Women ranchers—Texas—Matagorda County—Biography. 6. Hawkins family. 7. Matagorda County (Tex.)—History. I. Title. II. Series: Centennial series of the Association of Former Students, Texas A&M University ; no. 121.
F392.M4F87 2014
976.4′132—dc23
2013040905

Frontispiece: The Hawkins Ranch House viewed through the front yard gate, 2008. Photograph by J. R. Mullen, HRLTD

Illustrations with the notation "HRLTD" are courtesy of Hawkins Ranch Ltd.; all others are as credited in the captions.

To the persons and places of this book, to my extended family, to the "young lady ranchers," to the good people of the town and the country, and to the memory of my brother Frank Hawkins Lewis, for whom the Hawkins Ranch was a cherished place

Galleries of illustrations follow the genealogy and chapters 9, 21, and 27.

Genealogy
OF THE HAWKINSES, ALSTONS, AND RUGELEYS

HAWKINS GENEALOGY

For ease in tracing Hawkins Ranch family ancestry, key names appear in italics.

I. *Philemon Hawkins* the first (b. 1690 in Gloucestershire, England)
Founder of Hawkins family in America
Wife: Ann Eleanor Howard

II. *Philemon Hawkins* the second (September 28, 1717–1801)
Born in Virginia, moved to North Carolina; this Philemon is the subject of the 1829 oration "Colonel Philemon Hawkins, Sr." by his grandson John Davis Hawkins
Wife: Della Martin

III. *Philemon Hawkins Jr.*, the third (1752–1833)
Lived at Pleasant Hill in Warren County, North Carolina (First named Granville, then Bute, and subsequently Warren County)
Wife: Lucy Davis
Thirteen children, of whom two are listed here:
William Hawkins, elected governor of North Carolina in 1810, married Ann Swepson Boyd
John Davis Hawkins, married Jane Boyd

IV. *John Davis Hawkins* (April 15, 1781–December 5, 1858)
Born at his father's home, Pleasant Hill, in Warren County, North Carolina; lawyer and trustee of University of North Carolina; lived at his home, Spring Grove, in Franklin (now Vance) County, North Carolina; buried at Oakwood Cemetery near Raleigh, North Carolina

Wife: Jane A. Boyd, daughter of Alexander Boyd of Boydton, Mecklenburg County, Virginia

Eleven children:

James Boyd Hawkins (1813–1896), married *Ariella Adella Alston*
Moved to Texas

Frank Hawkins (1815–1896), married Ann Caroline Read
Moved to Mississippi

Dr. William J. Hawkins (1819–1894), married (first) Mary Alethea Clark, (second) Lucy N. Clark, (third) Mary A. White

John Davis Hawkins Jr. (1821–1902), married Ann Olivia Clark
Became partner of his brother James Boyd Hawkins in 1846 in a new plantation at Caney Creek, Matagorda County, Texas

Philemon Benjamin Hawkins (1823–1901), married a cousin, Fanny M. Hawkins

Dr. Alexander Boyd Hawkins (1825–1921), married Martha L. Bailey of Jefferson County, Florida, in 1858; Alexander (Sandy) Hawkins cultivated citrus in Florida and sent his brother James B. Hawkins orange trees, which Ariella planted at the Hawkins Ranch House

Ann Hawkins, married Col. Wesley Young

Lucy Hawkins, married Thomas Kean

Mary Frances Hawkins, married Protheus E. A. Jones

Virginia Boyd Hawkins, married William J. Andrews

Jane A. Hawkins, unmarried

V. *James Boyd Hawkins* (December 27, 1813–May 13, 1896)

Born at his father's house, Spring Grove, in Franklin (now Vance) County, North Carolina; moved to Texas in 1846 and established a sugar plantation on Caney Creek in partnership with his brother John D. Hawkins Jr.; phased out of sugar production and, with his son Frank Hawkins, began raising cattle about 1870; the story of the Hawkins Ranch in Texas begins with the James Boyd Hawkins sugar plantation and development of the cattle business following the Civil War; died at his Ranch House on Lake Austin in Matagorda County, Texas

Wife: *Ariella Adella Alston* (January 26, 1821–March 9, 1902)

Nine children (see appendix, Sketches and Letters of the Antebellum Children):

Sallie Hawkins (1837–1860), married Ferdinando Stith of Memphis

Willis Alston Hawkins (1839–1863), married Leah Irwin (or Erwin) of Warrenton, North Carolina

Virginia Hawkins (Jenny or Little Sis, 1840–1930), married Edmund Gholson Brodie of Henderson, North Carolina; their only child was James (Jimmy) Hawkins Brodie

James Boyd Hawkins Jr. (1843–1857)

John Davis Hawkins II (1844–1865)

Ariella (Ella) Hawkins (1846–1851)

Annie Hawkins, died in 1852 as an infant of three months; not mentioned in letters

Frank Hawkins (1847–1901), married *Elmore Rugeley Hawkins*

Edgar Hawkins (Charles Edgar Hawkins, 1849–1887), married Annie Lewis Hardeman; lived at and inherited the Hawkins plantation on Caney Creek.

- Four children:
- James Boyd Hawkins II, married Fannie Bruce
  - One child: Annie Dora Hawkins, married Mack Graves
- Frank Hawkins II (1882–1932), married Nettie Franks
  - Two children: Nettie Hawkins and Frank Hawkins III
- Edgar (Ned) Hawkins, married (first) Ruby Daugherty
  - One child: Savanna Hawkins, married Donald Duson
    - Married (second) Ruby Glasson, no children; when widowed, Ruby married Raleigh Sanborn (Raleigh and brother Camden Sanborn were sons of Annie Hardeman Hawkins and her second husband, John Sanborn); Raleigh and Ruby Sanborn operated part of the Caney plantation land as a farm and ranch and built a house in Bay City using the hand-made bricks of James B. Hawkins's sugar mill
- Ella Hawkins, married Seth Taylor
  - One child: Frances Taylor Steves, married Bert C. Steves
    - Five children: Marthella, Taylor, Lewis, Savanna, Diana

VI. *Frank Hawkins* (December 17, 1847–February 25, 1901)
Born in the town of Matagorda, died in Austin, Texas
Wife: *Elmore Rugeley Hawkins* (1867–1896)
Five children:
Henry Boyd Hawkins (Harry, August 21, 1888–December 9, 1951), unmarried
Meta Hawkins (March 11, 1891–January 7, 1975), married James Claire Lewis
Janie Hawkins (January 4–February 6, 1958), unmarried
Elizabeth Hawkins (Lizzie, January 20, 1894–March 3, 1957), divorced, no children
Elmore Hawkins (Sister, or Aunt Sister, April 3, 1896–March 7, 1975), married Esker L. McDonald, no children

VII. *Meta Hawkins Lewis* (March 11, 1891–January 7, 1975)
Married James Claire Lewis
Three children:
Frank Hawkins Lewis (January 11, 1920–June 6, 2003)
James Claire Lewis (J. C., 1925–1927), died at age two
Margaret Lewis Furse (b. October 25, 1928)

VIII. *Frank Hawkins Lewis* (January 11, 1920–June 6, 2003)
Returned from World War II to manage the Hawkins Ranch until sale of the cattle herd, April 1, 1999
Wife: Florence Neely
Four children:
Frank Hawkins Lewis Jr. (b. August 3, 1950), unmarried
Janet Lewis Peden (b. August 25, 1951), married Scott Peden
One child: John Hawkins Peden (b. April 14, 1983)
Meta Lewis Hausser (b. April 14, 1954), married Albert F. Hausser
One child: Albert Ford Hausser Jr. (b. May 20, 1987)
James Neely Lewis (b. July 21, 1956), unmarried

VIII. *Margaret Lewis Furse* (b. October 25, 1928)
Married Austen Henry Furse Jr.
Four children:

Jane Hawkins Furse (b. April 5, 1957), married John H. Friedman
Two children:
Elizabeth Hawkins Friedman (b. May 7, 1986)
Meredith Claire Friedman (b. July 30, 1991)
Austen Henry Furse III (b. June 6, 1960), married Anne Seel
Three children:
Katherine Burnett Furse (b. April 22, 1998)
Claire Lewis Furse (b. June 14, 2000)
Austen Henry Furse IV (b. March 23, 2003)
John Lewis Furse (b. February 3, 1962), married Susanne Nitter, no children
Mary Elmore Furse (b. July 21, 1965), married William J. McMillin
One child: Margaret Ann Eleanor McMillin (b. February 22, 2002)

ALSTON GENEALOGY

Willis Alston (1769–1837)
Born near Littleton, Halifax County, North Carolina; father of Ariella Alston Hawkins; attended Princeton, engaged in agriculture, and served in the Congress of the United States; his house near Littleton, North Carolina, was called Butterwood
Wife: *Sarah (Sallie) Madaline Potts Alston* (b. 1780); her daughter Ariella, son-in-law James B. Hawkins, and granddaughters wrote her many letters from Texas
Five children:
Charles Julian Poydroas Alston (b. March 12, 1818), married Mary Janet Clark
Ariella Adella Alston (January 26, 1821–March 9, 1902), married *James Boyd Hawkins*; nine children (see James Boyd Hawkins in preceding Hawkins genealogy)
Leonidas Alston (1823–January 27, 1849), married Emma Foster
Missouri F. Alston (Zuri, or Zury, b. November 2, 1824), married Archibald Davis Alston, son of Nicholas Faulcon Alston and Elizabeth Crawford Davis

Edgar Alston (April 20, 1827–September 8, 1848), died in Galveston

Rugeley Genealogy

Dr. Henry Lowndes Rugeley

Wife: Elizabeth Tabitha Elmore, married December 1, 1865

Nine children:

Elmore Rugeley (July 2, 1867–April 4, 1896), married *Frank Hawkins*

Five children: Henry (Harry) Boyd Hawkins, Meta, Janie, Elizabeth, Elmore (see Hawkins genealogy)

Ashton Rugeley (March 26, 1869–September 17, 1887)

Henry Rugeley (July 28, 1873–November 30, 1936), the "Uncle Henry" who at their father's death was appointed trustee for the five children of Elmore Rugeley and Frank Hawkins.

Eliza Colgin (December 17, 1875–May 17, 1876)

Elizabeth Fitzpatrick (March 3, 1877–October 14, 1877)

Sue Lewis (Dolly, b. December 15, 1878), married Otikar (Otie) Doubek.

One child: Jane Doubek Gainer

Philip Ludlow (Lud, December 12, 1881–1905)

Edith May (January 6, 1884–November 19, 1895)

Rowland (February 19, 1889–April 19, 1979), married Lenore Wall (Daughty)

One child: Martha Rugeley Bachman, married Richard C. Bachman

Two children: Rowland Rugeley Bachman and Richard Cloar Bachman Jr.

Repetition in First Names

Philemon

Three Philemons in succession: the father, grandfather, and great-grandfather of John Davis Hawkins; a fourth, Philemon Benjamin, was the son of John Davis Hawkins

John

John Davis Hawkins, father of James B. Hawkins

John Davis Hawkins Jr., son of John Davis Hawkins; partner and brother of James B. Hawkins

John Hawkins, son of James B. Hawkins

Willis

Willis Alston, father of Ariella Alston Hawkins, her sister Missouri, and their three brothers

Willis Hawkins, son of Ariella and James B. Hawkins

Willis Walter Alston, son of Missouri and Archibald Davis Alston

James

James Boyd Hawkins, son of John Davis Hawkins, husband of Ariella Alston

James Boyd Hawkins, son of Edgar Hawkins, grandson of James Boyd Hawkins and Ariella

James Boyd Hawkins, son of the preceding James, grandson of Edgar Hawkins

Sallie

Sarah (Sallie) Alston, mother of Ariella Alston Hawkins

Sallie Hawkins, daughter of Ariella and James B. Hawkins

Frank

Frank Hawkins, who settled in Mississippi, second son of John Davis Hawkins, brother of James B. Hawkins

Frank Hawkins, son of Ariella and James B. Hawkins, namesake of his Mississippi uncle; married Elmore Rugeley; father of Henry (Harry), Meta, Janie, Elizabeth (Lizzie), and Elmore (Sister) Hawkins

Frank Hawkins II, son of Edgar Hawkins, grandson of James B. Hawkins

Frank Hawkins III, son of Frank Hawkins II, grandson of Edgar Hawkins

Frank Hawkins Lewis, son of Meta Hawkins and James Claire Lewis, grandson of Frank Hawkins

Frank Hawkins Lewis Jr., son of Frank Hawkins Lewis and Florence Neely Lewis

Edgar

Edgar Alston, younger brother of Ariella Alston Hawkins; died at age twenty-one in Galveston

Edgar Hawkins (exactly, Charles Edgar Hawkins), second Texas-born son of Ariella and James B. Hawkins; he and his family inherited the Caney plantation; upon his early death his widow married John Sanborn

Edgar (Ned) Hawkins, son of the preceding Edgar Hawkins

Elmore

Elmore Rugeley Hawkins, daughter of Dr. and Mrs. Henry Lowndes Rugeley, wife of Frank Hawkins, mother of the five children born at the Hawkins Ranch House; her death at the birth of her youngest child resulted in her young children being cared for by her parents in Bay City, Texas; Elmore was the maiden name of her mother, Elizabeth Elmore Rugeley

Elmore Rugeley Hawkins (Sister), youngest of the five children of Elmore Rugeley and Frank Hawkins, married Esker L. McDonald

Ariella and Ella

Ariella Alston Hawkins, wife of James Boyd Hawkins; frequently called Ella, but Ariella has been used throughout this book

Ariella (Ella), daughter of Ariella and James B. Hawkins, died as a young child

Ella Hawkins Taylor, daughter of Edgar Hawkins, sister of James, Frank, and Ned Hawkins; married Seth Taylor, one child: Frances Taylor Steves

The Hawkins Ranch in Texas

Hawkins Ranch in Matagorda County

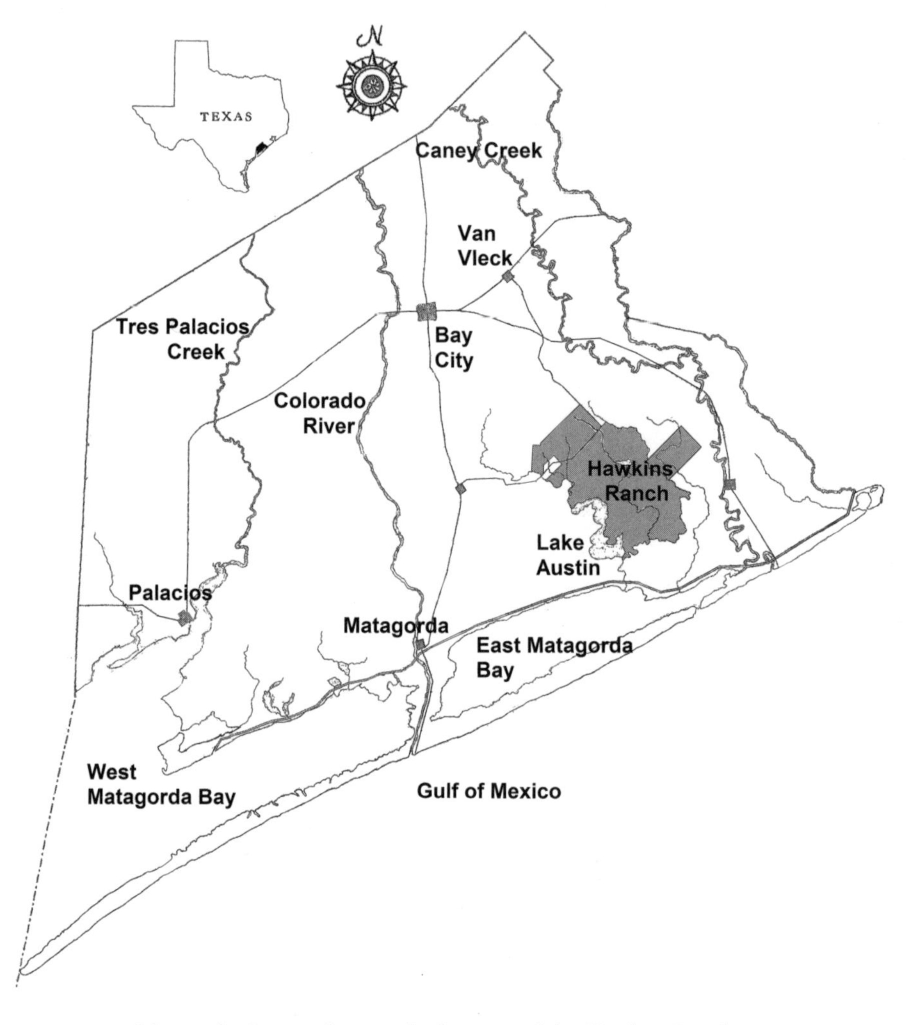

Matagorda County showing the location of the Hawkins Ranch in the southeast, between Caney Creek and Lake Austin. The silted area dividing Matagorda Bay at the Colorado River mouth was not present in J. B. Hawkins's day. Adapted from a county map at the Matagorda County Museum, used by permission

Boundaries of the Hawkins Ranch, showing Lake Austin with the Hawkins Ranch House beside it and, to the east, the remains of the Hawkins plantation. Farm Road 521 was present only after 1945. Map courtesy Bill Isaacson, HRLTD

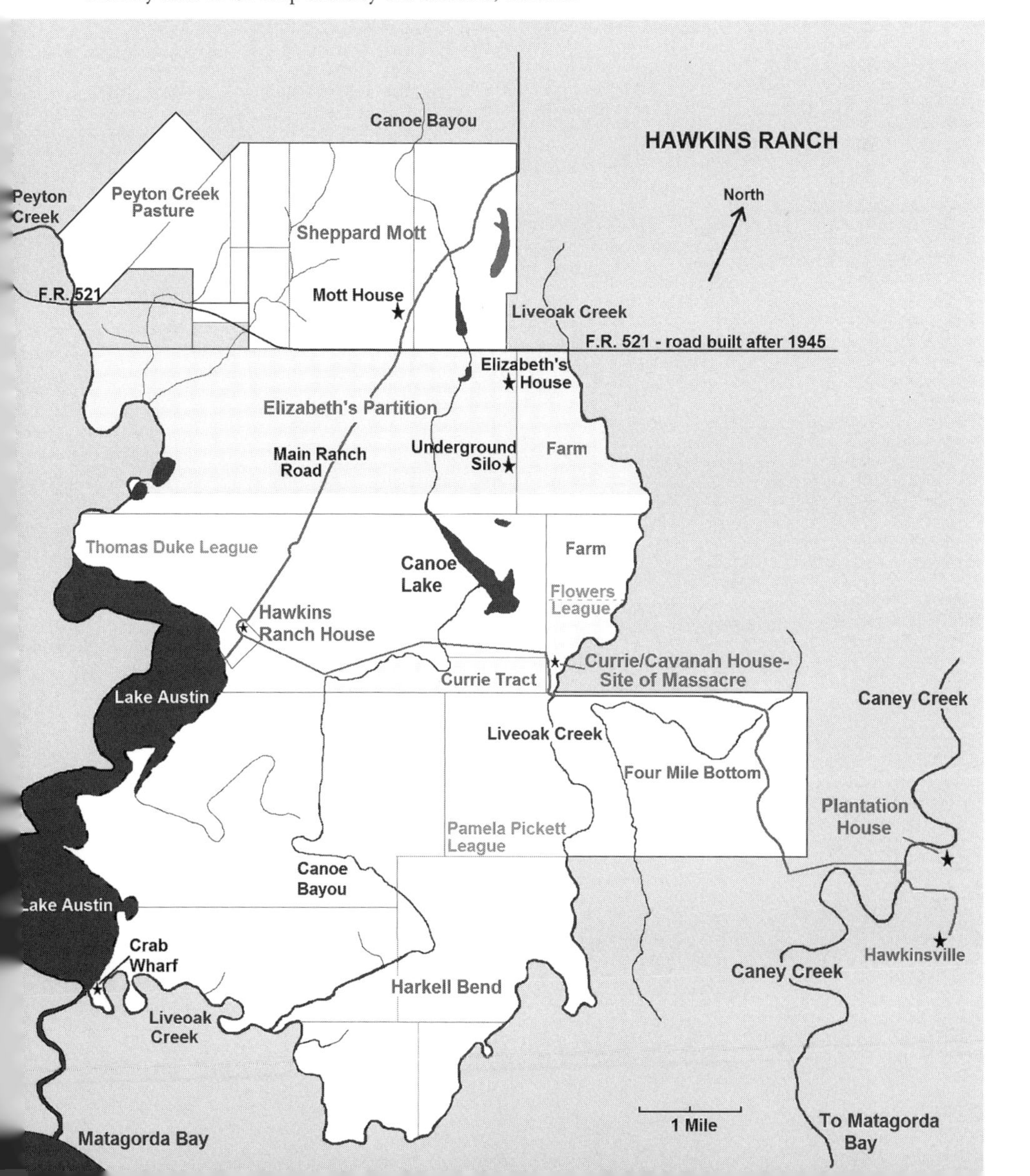

Introduction

In 1936 teachers in Texas used every means they could imagine to impress children with the importance of that year. It was the Texas Centennial, the hundredth anniversary of the independence of Texas from Mexico. From Jefferson Davis Grammar School in Bay City, we were marched two by two to the City Auditorium, with my first grade teacher Miss Tenie and others patrolling. There we sang the official state song that we had rehearsed, "Texas, Our Texas! All Hail the Mighty State!" Appropriate school and city dignitaries listened and spoke. Back at our classroom tables, using small, blunt scissors, we cut out the traced outline of Texas.

We saw how its rectangular panhandle pushed into the border of Oklahoma, how the "Big Bend" of the Rio Grande drooped like an elephant ear into Mexico, and how the coast of the Gulf of Mexico, where we went to the beach on Sundays, made a gentle crescent. The Red River made an irregularly chiseled border with Oklahoma, as did the Sabine River with Louisiana. We were asked to consider where we ourselves were located, and it turned out to be in Matagorda County, close to the Gulf Coast, between the Brazos and Colorado rivers.

This was always the location of the Hawkins Ranch families, whether they were in town or country, or whether the date was their arrival about 1846 or many decades afterward. *Place* was of central importance to the Hawkinses in every decade of their experience, and place included broad coastal acres, the Hawkins Ranch House itself, and, once it was established, the town of Bay City. James B. Hawkins and his wife Ariella, my mother's grandparents, were the antebellum pioneer generation who "set forth." In the century and a half that followed, the next Hawkins generations, including my mother's, "settled in." This story is about the Hawkins family who came from North Carolina, settled in one Texas location, and stayed for five generations. The changes that impacted their lives derived from changing times, attitudes, and social assumptions—not changing locations.

The coastal region where J. B. and Ariella Hawkins first established a sugar plantation and then a cattle ranch included leagues colonized by

Stephen F. Austin beginning about 1822. The Texas Hawkinses were not original Austin colonists, but the leagues within land they acquired bore the names of Austin's colonists: Pickett, Duke, Defea, Dinsmore, Dwyer, McCarty, Fry, and Byrne. The Hawkins Ranch House was and is located within the league originally assigned to Thomas M. Duke, one of Austin's old three hundred who served as an alcalde under him. There is some indication of kinship between the Hawkins Ranch family and Joseph H. Hawkins, who came to the aid of Stephen F. Austin in New Orleans, but neither Stephen F. Austin nor Joseph H. Hawkins is mentioned in J. B. Hawkins's letters.[1]

After the Civil War and Emancipation, my mother's father, Frank Hawkins, the first Texas-born child of J. B. and Ariella, helped his parents change from planting and sugar production to raising cattle. Beginning in 1917 my mother, Meta Hawkins Lewis, and her sisters continued their father's cattle business for three decades. Having spent many days and years in their company, I have been able to describe the way they took over the management of their ranch, their deliberation about the fate of their then-collapsing Ranch House, and their grave concern over the separation from them of their sister, Lizzie. From my firsthand experience I have also described the post–World War II management of the Hawkins Ranch by my brother Frank Lewis. A brief outline of the current management by the younger generation of partners does not so much conclude the story as it foretells the question for the future: Will *place,* in the sense of location, in the sense of the town of Bay City, and especially of the house and acres of the Hawkins Ranch itself, continue to be the keenly-felt mooring of a family if it has scattered to many places? Or when the assets to be managed include impersonal ones like securities that, unlike a *place,* do not compel allegiance?

My Main Sources

For the antebellum Hawkinses and their plantation life, I have used letters sent by J. B. and Ariella Hawkins from their Texas plantation to relatives in North Carolina from 1846 to 1860. These letters are archived in the Southern Historical Collection of the Wilson Library at the University of North Carolina at Chapel Hill.[2]

I learned of these collected letters through my nephew, Frank H. Lewis Jr., and his correspondence with a cousin, Tim Hawkins of Concord, North Carolina. The letters sent from the sugar plantation are rich in details of what children were doing, how the garden and the commercial crops were faring, who was suffering "chills and fever," and what remedies were dispensed for their cure. Rather than paraphrasing these letters extensively, I have often quoted them in order to bring alive the first-person experience of the writers and show their use of language and the subjects that interested them. Occasional idiosyncratic spellings have been kept. The early Hawkins family in Texas was greatly concerned to stay connected with their North Carolina relatives. Writing and receiving letters was enormously important to them. They pleaded for letters and were apologetic if their own replies were late. In 1847 Ariella apologized for being late in replying to her mother by explaining that she was "prevented by the birth of a fine boy," who was Frank Hawkins, my mother's father.

Helping me examine the letters at the University of North Carolina in May 2006 were my son John Furse and granddaughter Elizabeth Friedman. As we looked through the fragile letters, the reading room was silent except for an occasional cough, a whisper, or the squeak of a tennis shoe. Then John's eye fell on a letter dated March 22, 1854, and he let out a whoop that lifted all the studiously bowed heads in the room. The letter filled us with excitement, because for the first time we found out the date when the Hawkins Ranch House was built: 1854. My mother and her siblings, although born in the house in the late 1800s, had never known the date or details of its building. J. B. Hawkins wrote to his mother-in-law Sarah Alston that he was then "busy sawing out lumber for the Lake Aston [Austin] House. She [Ariella] is going to put up a large and splendid building." He added that when it was finished, "I hope . . . to have you to live with us. I think we will make a pretty place of it."[3]

The antebellum letters give a picture of a slave-worked sugar plantation on the banks of Caney Creek in Matagorda County, Texas, but they omit matters we would like to find in them. The family seemed to live as if the plantation, a dwelling and also a busy workplace, was a world in itself. They do not report on public events of the county, state, or nation. Nor do they refer to books, newspapers, or current events. J. B. and Ariella Hawkins arrived in Texas in 1846, only one year after Texas gained statehood on March 2, 1845, following a nine-year period (1836–45) as an independent republic, but these letters from the plantation in Texas have nothing to say about the transition. There is no reference to Stephen F. Austin and his colonists, who came to the same region about twenty-four years earlier. Ariella's younger brother Edgar Alston served in the Mexican War and returned to live in Galveston, but the war itself is mentioned only as it relates to Edgar's serious financial difficulties. The letters cease, as would be expected, just as the Civil War begins. The experience of the Hawkins family during the Civil War can only be pieced together through sources other than these letters. That J. B. Hawkins sold goods to the Confederacy and also helped with Confederate coastal defenses near his plantation are facts known through other sources, including his personal notations and receipts.

My second main source for this family social history is my memory, beginning in the 1930s, of the family, the town, the ranch, and the region. My experience has been reinforced by excellent local historical records, news clippings, the memories of others, and the private Hawkins Ranch File of Historical Data. This file is located at the office of the Hawkins Ranch (2020 Avenue H, Bay City, Texas 77414). It includes a few letters and sales receipts for barrels of sugar and molasses, some of which were sent to the Confederacy's "Department of Texas." The file includes pocket memorandum books of J. B. Hawkins, a few documents relating to the generation of my mother and her sisters, their management of the ranch, and the building in 1913 of their red-tile-roofed house in the town of Bay City.

The plantation land and house on Caney Creek were willed to Frank Hawkins's younger brother Edgar Hawkins and his family, who, after many years, sold it. This plantation house where J. B. and Ariella first settled, purchased from the Quick family, fell into disrepair and finally collapsed. Its remains are still identifiable at Hawkinsville.

The Hawkins Ranch House, built in 1854 by J. B. Hawkins on Lake Austin, became the home of Frank Hawkins and his wife, Elmore Rugeley Hawkins, after their marriage in 1887. It was the headquarters of the ranch that Frank Hawkins operated and the birthplace of his five children—Henry Boyd Hawkins (called Harry), Meta Hawkins Lewis (my mother), Janie Hawkins, Elizabeth Hawkins (Lizzie), and Elmore Hawkins McDonald (called Sister). As I knew them well, I introduce them in part II, Young Lady Ranchers, in my own voice and from my experience.

I have sought to tell this story through personal characterizations and episodes rather than as a fluid narrative. Often the story turns on character traits like the optimistic confidence and hard work of J. B. and Ariella Hawkins as they began a plantation in Texas and like the courage of the "young lady ranchers" in taking over the management of their cattle ranch from their more experienced uncle. The story also shows how psychological balance or lack of it can have very practical consequences. The fierce instability of Ariella (in her last days) and of Lizzie (in the 1930s and 1940s) risked not only the loss of family love and loyalty but the loss of thousands of acres.

This family history is not the product of the kind of research aimed at relating the story to public events that become the story's frame and rationale. However, the two principal sources I have used allow close observation of social change and changing assumptions over an extended period and therefore invite historical interest. It is rare to have such detailed access to a family's domestic life as the two sources provide.

Mainly this story shows family life lived in the same place for more than a century and a half. Someone has said of history that it consists of the record of what men have done in the out-of-doors: armies march, the kingdom falls, and there is a new date for children to memorize. Unrecorded goes the important fact that a small child was rocked in a cold room in the dead of the night. The vital details of family life—illnesses, schooling, quarrels, planning, worries, failures, recreations—tend to be saved in impermanent letters and fleeting conversations. But just this part of family life has been especially accessible to me in the two sources I have used. An excellent book that fills out the picture of the region and its people beyond a single family's story is the Matagorda County Historical Commission's *Historic Matagorda County.* It has contributions by people of the region and was researched, organized, and edited by the late Mary Belle Ingram and by Frances Vaughn Parker, who also created its historical maps.

Three aspects of this story give it wider significance and interest. It provides glimpses of the past through the lens of one family. It also provides some practical and psychological clues for family business in general, especially the ranching business. And it illustrates two ways in which a family commonly relates to its land: the entrepreneurial way that regards land as a business opportunity, and the way of emotional attachment that sees it as a received heritage.

Glimpses of the Past through the Lens of a Family

Coming into view are issues like slavery and its aftermath, education, the role of women, livelihood and leisure, manners and home life, modes of transportation, town and country life, property rights, architecture and materials, illness and medicine. I will select three of them for illustrative comment.

Slavery. So uncritically accepted by the slaveholding planters and so abhorrent to the current writer and reader, slavery is an arresting issue early in this story. I need to forewarn the reader that the view we are given of slavery comes from slave holders' letters and therefore does not represent a larger picture or the expressed attitudes of the slaves themselves. Other sources giving a fuller picture of slavery and the experience of African Americans in Texas are *Till Freedom Cried Out: Memories of Texas Slave Life,* edited by T. Lindsay Baker and Julie P. Baker; Randolph Campbell's *An Empire for Slavery: The Peculiar Institution in Texas 1821–1865;* Alwyn Barr's *Black Texans: A History of African Americans in Texas 1528–1995;* and *Black Cowboys of Texas,* edited by Sara R. Massey with an introduction by Barr.

In establishing a plantation in Texas, the Hawkinses brought with them not only slaves but a *system* of farming that presupposed slavery as the accepted norm. This family record gives a view of how they and others like them managed a slave workforce. The Hawkinses already owned slaves and came to Texas because they needed more and cheaper land for them to work. In their letters they do not clearly state this motive, but it is a justifiable inference from the letters they write of their plans to acquire more land in order to keep their workforce working. It is unlikely that they would offer a clear explanation in letters to family members who already understand their motive in leaving.

Because farming with slave labor was a system, and one the planters simply assumed and operated under, the Hawkinses did not attain any perspective from which to see slavery as reprehensible. The circumstance of Edgar Alston, Ariella's brother, who owned slaves but did not choose to farm, gives some insight into the way slave holders thought. Because they saw themselves as benevolent, they appeared to think their responsibility was discharged by the good care they assumed they gave their slaves. They vigorously condemned "poor management" and neglect of care. On J. B. Hawkins's arrival in Texas, he was outraged to discover that his temporary overseer had failed to house and care for his slaves as instructed. He found many sick when he arrived. The Hawkinses simply brought the plantation slave system with them from North Carolina, were already familiar with it, accepted it, and made no comment upon it beyond the practicalities of workforce management. After the Civil War and the end of slavery, the Hawkinses turned toward ranching.

The Role of Women. I was surprised to see how much the role of women came into play in the Hawkins story. Ariella, the wife of James B. Hawkins, who arrived in Texas with him and their children, is a central figure. Because most of the letters written from the plantation in Texas are written by her, we are given a close view of the home life of the plantation, the children's games, and the leisure activities of the family, which involved horseback riding and visits to other sugar plantations. She is a good seamstress and gardener and, as a letter writer, an excellent describer of the scene. She is generally in charge of treating the frequent illnesses ("chills and fever"—the symptoms of malaria) that plagued both her own family and the plantation slave families. Late in her life, after her husband's death, Ariella had a spirited quarrel with her son Frank Hawkins because her community property rights were overlooked in her husband's will, a circumstance that favored her son Frank. She was intelligent in knowing what her rights were, energetic in locating a lawyer to represent her, and vigorous in arguing her case. But as age and anxiety impaired her judgment, she veered far off the course of reasonableness. Her son made amends and paid her for the rights of which she had been deprived. Had her suit not been settled, the heirs of Ariella's lawyers might now own part of the Hawkins Ranch because they agreed to accept their fee in a percentage of the land they expected to recover for her.

I have called my mother and her sisters the young lady ranchers, because in 1917, when still in their twenties, they took over the management of the Hawkins Ranch from their uncle and trustee, Henry Rugeley. Their parents both died while they were small children, and their uncle probably would have continued running the ranch for them indefinitely, thinking that these inexperienced young women were not capable of managing it themselves. My mother and her sisters did not see themselves as feminist pioneers. Indeed, the concept was not within their reach. But they read the will of their father, Frank Hawkins, and he had plainly left them their ranch free of trust when the youngest of them turned twenty-one. When that day came on April 3, 1917, they went to their uncle and in the politest way possible told him they would now take over the management of their ranch. For the next three decades (until my brother, Frank Lewis, began to manage it after World War II) they ran their ranch of about thirty thousand acres.

Only four years later, on March 31, 1921, Lizzie unexpectedly asked her sisters to partition to her the individual interest she held in the Hawkins Ranch, so that she could farm and ranch it independently of them. Her attempt at independence was coupled with grand theatrical gestures and ended in utter failure. After many years and at great financial and emotional cost, Lizzie's sisters purchased her partitioned land from her creditors and brought it back into the acreage held by the Hawkins Ranch partnership. Sadly, she was estranged from her sisters until her death on March 3, 1957.

Town and Country. J. B. Hawkins stationed his wife and children for months at a time in the town of Matagorda. Theirs was a relationship to town and country that meant renting a house in town and transporting a group of servants and household items by wagon for a long stay. In the lives of my mother and her sisters, beginning in the 1920s, the automobile became the means by which ranchers could live in town (now Bay City instead of Matagorda) and daily attend to their land and cattle in the country. Town and country, now less separated, were destinations in a shuttling automobile that could easily make a round trip in an hour, although a flat tire or getting stuck in the mud were frequent hazards.

My childhood observations of country life were made from the back seat of an automobile. When I arrived at the ranch I rode a horse and was in the company of the good African American "people on the place" who made up the farmers, cowhands, and cooks who worked at the ranch and sometimes in town. I knew them well and introduce them in Part III, The Instruction of Town and Country.

Town life of the 1930s I observed by walking, usually barefoot, on sun-baked sidewalks. From childhood memory I am able to give a picture of a small town with a population of about six thousand and its businesses that operated in store fronts surrounding the courthouse. In the 1930s in the small town of Bay City, *a business was a person.* Almost always the proprietor was there in the place of business and working behind the counter. There were no absentee owners, no franchises, and few brand names. Business was never so brisk that customers had to wait in a line or take a numbered ticket. A business transaction was a welcome few minutes of sociability, and social skill was regarded as a great business advantage. A stony indifference to entering into the social aspects of a transaction would elicit the complaint, "He won't even pass the time of day with you." Almost every business advertised itself with its owner's name, and it seemed to me that everyone in the town knew everyone else.

Operational Clues for a Ranching Family

Three main strategies have helped hold together this long-standing family business, the Hawkins Ranch partnership. The first has been to make the livelihood of the family visible to all family partners. Visibility of the family livelihood was a simple matter in J. B. Hawkins's day, when his wife and even his children could look at his fields and know the state of his crops and the expected yield, conditions they often reported in their letters.

Making commonly held assets visible becomes more problematic in modern times if the asset class begins to change from cattle and crops to the management of land, minerals, and securities, as is happening in the Hawkins Ranch partnership today. There is an abstraction in assets like these that requires interpretation and analysis. Today Hawkins Ranch holdings can no longer be indicated by pointing to pasture land and rows of crops. Visibility of the family livelihood begins to require a special effort at clarification for the sake of the partners. Making such an effort, through partners' meetings and clearly presented data, has been a great benefit to the current Hawkins family partnership, as I describe in the final chapter.[4]

The second strategy is to move from a natural paternalism of the elders of the family business to a collegial method of management that includes previously dependent partners. There might be resistance to such a move from a well-intentioned head of family who unwittingly confuses care with control. A legal instrument that provides for their inclusion is clarifying. The will of their father, Frank Hawkins, cleared the way for my mother and her sisters and allowed them to take over from their uncle.

The third strategy is to have a written partnership agreement that makes doubly clear what the common family partnership interests and responsibilities really are. J. B. Hawkins and his brother and partner John D. Hawkins Jr. signed such an agreement in October 1846, and 148 years later, on December 23, 1994, the current Hawkins Ranch partners signed a partnership agreement establishing Hawkins Ranch Ltd. A special advantage of the current agreement is that this makes it possible to sell or buy an interest in terms of a percentage of partnership interest rather than by partitioning actual land.

Two Ways of Relating to Land

Two identifiable attitudes toward land can be discerned in the Hawkins family history. J. B. Hawkins thought of the land he acquired in Texas as a practical opportunity to which no sentiment was attached. The attachment of sentiment may, in fact, require the passage of time. J. B. Hawkins's feeling toward his plantation and the projects that he oversaw was a feeling of pride in his own achievement. His holdings were not an inheritance valued because they derived from the past. J. B. Hawkins came to Texas and bought land for the practical reason of putting it to use as a place to plant crops, and if the practicalities had indicated this, he might have sold some or all of it without heartfelt regret or a sense of betraying the past.

Later generations, my mother's and those that followed, admittedly thought of the land as a way to raise cattle and earn a living, but great sentiment also came to attach to it. They saw their received place, especially the Ranch House and the surrounding land, as a *heritage* to be kept intact and changed as little as possible. For a member of the family to ask for a division of the property, as my mother's sister Lizzie did, seemed a betrayal. In 1935 the sisters' quandary about whether to allow the Hawkins Ranch House to collapse from old age was made difficult by the tension between utility and heritage. The house, although not then in use, was part of their heritage, but was its value as a heritage important enough to pay the practical cost of saving it?

The same distinction in ways of regarding land is made by Sonya Salamon in *Prairie Patrimony.* She examined the two cultural meanings of land for farmers in the Midwest. Using the tools of anthropology, she identified "yeoman German farmers," who are attached to land as a heritage, versus "Yankee entrepreneurs," who regard land unsentimentally as an investment. However, she ties each pole to an ethnic group or "community."[5]

What happens in a family business, especially a land-owning business like ranching, is that both poles influence the same group. As a family, people develop attachment to the place; yet as a business, it is subject to utilitarian detachment. Today's Hawkins Ranch partners continue to negotiate the tension between these two ways of regarding their land.

The recent history of the Hawkins partnership has been characterized by a diversification of its assets in which its land holdings have been augmented by other investments to which a commercial value but no sentimental value attaches. If the holdings of the Hawkins Ranch partnership become less centered in an inherited place and more centered in impersonal investments, then the partnership will likely conduct its business on almost purely utilitarian and practical lines. Such a shift raises the question of what, without a revered place, could produce force enough to bind the family to one another. Would the family become joined only in that remote and impersonal way that those who own the same stock or mutual fund are joined? If so, no allegiance would continue to inhabit the common interest, and any allegiance that does continue would derive from kinship alone.

The Plantation Place

J. B. and Ariella Hawkins came to Texas in a hopeful, expectant spirit. In letters they told of their Mississippi River boat trip and how they were setting up their plantation on Caney Creek in coastal Texas. This location, later called Hawkinsville, was the place of the plantation's beginning in 1846.

Almost a hundred years later, as a school girl in the 1940s, I sometimes came to this place with my parents on one of their frequent afternoon automobile trips to country places. They came here in order to visit the small fenced-off family cemetery located a few steps from the collapsing structure that had been the plantation house. J. B., Ariella, and most of their children were buried there. During her long life, Ariella outlived and saw buried her husband and all her children except her daughter Virginia.

My mother came to this place not so much in reverence to sense the air of "solemn stillness" that Gray evoked in his *Elegy,* although she certainly knew those lines that put a "lowing herd" and "mould'ring" graves in proximity, as was the scene before us here. She came to see if the fellow hired for the task had cut back the weeds around the grave stones as promised. The small family cemetery was now on land the Hawkins family no longer owned, but maintaining the graves was still their responsibility. A few yards from the cemetery we could see the remains of the house. The fields surrounding it had been the focus of J. B. Hawkins's very great energy and ambition.

Now, in the 1940s, a century after he arrived with Ariella and their children, this place had become a lifeless ruin with a story it had no means of telling. But no visitor could come here without wishing to hear the story's beginning, which, thanks to letters from those who first came, can now be told in their words.

Part I

Plantation Beginnings

According to Hawkins letters, the plantation established on Caney Creek was a busy workplace, not one of lazy self-indulgence for the planter and his family, as the word plantation may imply. J. B. Hawkins's activities were not limited to the supervision of planting by his slaves. He took orders for his products (sugar, molasses, cotton, hides, and lumber) and arranged shipping to his customers. J. B. Hawkins described himself as a very busy man and called his activities his "business." Beginning in 1846, the slave work force numbered 34; by 1860 it was 101. The main commercial crop, sugar cane, was labor-intensive. It was cut by hand and shoved a few stalks at a time into a grinding machine powered by a circling mule. The juice was directed by a metal lip into a bucket, poured into big iron kettles, and boiled down to molasses or sugar. Barrels and hogsheads of the product were shipped by boat down Caney Creek, ultimately to customers along Matagorda Bay. Between 1850 and 1855 a mile-long canal from Caney Creek to Matagorda Bay was dredged, permitting boat access to the bay. Before that time cargo had to be unloaded from the creek's southernmost point, hauled overland to the bay, and then reloaded onto boats.[1]

Topics addressed in Part I include establishment of the sugar plantation, details of family life, and construction of the Hawkins Ranch House; slavery, its ending, and the post–Civil War change from planting to ranching; and how the quarrel between Ariella Hawkins and her son Frank was resolved.

Chapter 1
NORTH CAROLINA ROOTS

In 1829 John Davis Hawkins was feeling that worrisome shudder that so often afflicts members of an older generation when looking at the young and finding them undisciplined. Hawkins was a North Carolina lawyer and planter, an alumnus of the University of North Carolina, and for fifty years one of its trustees.

His eldest son, James Boyd Hawkins, a boy of sixteen at the time, whose life would be lived out in Texas, was approaching college age. Should this son attend his father's revered alma mater, the University of North Carolina? The father had grave doubts and confided them in a letter to his classmate John Branch, the secretary of the navy. He told Branch that the students were too little disciplined by the faculty and that they were more given to "extravagant dissipation and drunkenness" than to "emulating each other for literary form." John D. Hawkins wrote this gloomy evaluation and said it made him think of West Point as perhaps a better alternative. But first he would like to make a visit there himself to get more information on West Point educational methods. Could John Branch, as secretary of the navy, possibly arrange to have him appointed an official visitor there the following June?[1]

James Boyd Hawkins (1813–1896) indeed enrolled at West Point as a member of the class of 1836. If careful planning, hard work, and diligence were inculcated at West Point, its influence on him was certainly apparent from 1846 until his death in 1896, in the way he set up and operated his Texas plantation and later cattle ranch.

At the very time of John D. Hawkins's worry about the youth of the day, his own father, Philemon Hawkins (the third of that name; see Hawkins genealogy) was also worried: that the younger Hawkins generation would not know or appreciate the fine character traits of his own father, Philemon the second. On September 29, 1829, at his father's request, John D. Hawkins invited family and friends to come to hear a finely-crafted tribute that he wrote and delivered. The speech was twice published as a pamphlet, first in 1829 and again by his youngest son, Dr. Alexander Hawkins, in 1906. In it John D. Hawkins tells a captivating story in the young life of the heralded Philemon (see endnote).[2]

John D. Hawkins praised his grandfather mainly for his work ethic. That was what the young should emulate, he said: hard work, avoidance of debt, attending to crops and to business. These habits and virtues would be rewarded by the accumulation of worldly goods, which would make one more useful to others.

Having had no education himself, Philemon the second had greatly valued it for his own sons, two of whom he sent to Princeton: Benjamin and Joseph Hawkins. Benjamin (1754–1818), the uncle of John D. Hawkins, became so adept in French that he was asked to serve Gen. George Washington as a translator for French troops during the American Revolution.[3]

In the vigorous ending of his speech, John D. Hawkins forcefully impressed upon his audience the importance of his grandfather's character and example:

> When we take a review of his rise and progress in life, and contrast them with the idleness and dissipation of the present day, we are ready to exclaim, that degeneracy is surely among us. He lived within his income, and caused it continually to increase; by which he was not only increasing his ability to live, but to increase his fortune, and to add to his power to be useful. . . . Let us then emulate his virtues, inculcate his habits, and instill into the minds of our children the examples of his prosperous and useful life; and when each rolling year shall bring around the day of his birth, let us hail it as his natal day, and endeavor to imprint it deeper and deeper in their hearts.[4]

James B. Hawkins did not stay at West Point to graduate and never embarked on a military career. Instead he married Ariella Alston in 1836. After their marriage at her father's residence, Butterwood, near Littleton in Halifax County, North Carolina, the couple spent the next ten years living in North Carolina, in close touch with their respective families. Both the Hawkins and the Alston families were planters and used a slave workforce. In North Carolina the extended Hawkins family had various business pursuits related to agriculture—milling, potash production, cotton ginning, and tobacco—and eventually a primary interest in the development of railroads. J. B. Hawkins was involved in planting and in the various business pursuits in which his father was engaged. Ariella was busy with her growing family. She and her husband had six North Carolina–born children who went with them to Texas, where two more were born.[5]

Ariella's father, Willis Alston (1769–1837), died shortly after his daughter's marriage; most family letters from Texas starting in 1846 were written to her widowed mother, Sarah Madaline Potts Alston. She was constantly urged to come to Texas for a visit but probably never did. Ariella's father, Willis Alston, attended Princeton and served in the United States Congress during the presidency of Thomas Jefferson. He was a member of Congress for several terms (1803–15; 1825–31) and during his service in 1812 was chairman of the Committee on Ways and Means. Ariella's mother, called Sallie, was his second wife but the only one by whom he had children. Altogether, there were five Alston children, most of whom were addressed or mentioned in letters from Texas.[6]

It was a brave undertaking for James B. Hawkins and Ariella to leave North Carolina and their extensive Hawkins and Alston families and move with their many small children to the largely unsettled place that Texas then was. Earlier, in 1844, in what was apparently a staging operation, James and his brother Frank had traveled together from North Carolina to Mississippi. There J. B. Hawkins left some of his slaves to be retrieved and moved to Texas later. In 1848 Frank settled permanently in Mississippi.

Frank probably chose Mississippi because he intended to farm cotton. But J. B. Hawkins had become enthusiastic about growing sugar cane and producing molasses and sugar, as was being done among growers along Caney Creek. Land around Caney Creek in Texas was fertile, and the creek itself provided a means of getting sugar to market. Ariella's brother Edgar Alston said Caney Creek in 1845 was navigable from a point above the land that J. B. Hawkins and his brother John D. Hawkins Jr. had bought. Other sugar growers, a number of whom were from North Carolina, had also settled along Caney Creek. In 1847 J. B. Hawkins wrote that many planters along Caney "have quit cotton and are going altogether on sugar. They are putting up splendid brick sugar houses. There are 4 large sugar houses within 2 miles of each other. I examined all the crops. They were splendid."[7]

Although James and Frank left North Carolina and settled in different states, their motive for leaving home was the same. It was primarily the fact of slave ownership. A planter already having a slave labor force was highly motivated to have land enough for such a force to cultivate. Slave holders regarded such a force as a business asset that could not simply lie idle. If moving was required in order to acquire enough land to make use of the existing workforce, then the families moved to new territory.

Chapter 2

Letters Written en Route

Traveling to Texas, Ariella wrote to her mother on November 20, 1846, when she arrived in New Orleans from Memphis. The fact that New Orleans was a new and thrilling adventure for her suggests that Ariella probably had not been to Texas before the trip she now described.

She was "heartedly tired of traveling" and anxious to get to their Caney Creek home in Texas. Traveling in a paddle wheeler, they had started down the Mississippi from Memphis. Probably they arrived at Memphis from North Carolina by private carriages or public stage. The slaves, who accompanied them on the Mississippi riverboat, had made their way from North Carolina to Memphis in their own group, probably by wagon. All reached Memphis safely, Ariella told her mother, who she knew would want to hear, and were "well pleased" with their trip.[1]

The party traveling to Texas together consisted of Ariella and her husband, whom she called "Mr. Hawkins"; their six children, including their baby daughter Ella; and the wife of the (probably permanent) overseer hired for their Caney plantation. Ariella did not say whether the overseer himself was on board. While J. B. Hawkins was bringing his family to Texas, he had put a temporary man named McNeel in charge at Caney. The trip from Memphis to New Orleans took them one week by paddle wheeler.

The Mississippi River was low, with so many hazardous snags that the captain decided it was not safe to run at night when the obstacles could not easily be seen. Ariella described these dangers:

> We passed five boats sunk between Memphis and New Orleans, two of them very large boats, were lost a few days before we reached Memphis. [She may mean a few days out of Memphis or a few days before reaching New Orleans.] We passed the wreck of the Monarch just at night and stopped close along side of her, so close as to run against her and break a shaft and two

> arms of one of our wheels. It made a tremendous noise when we run against her and we had quite a scare on board. Two ladies fainted; one of them was our overseer's wife. I thought at the time she would never come too[,] she was so dredfully frightened. We then run upon one wheel over to the wreck of another very large boat the Matamoros, took off some of her fraight[,] run a few miles down the river and stopped for the night.[2]

Their adventures were not ended. Their sleep was disturbed about midnight, and Ariella "was awakened by the cry of fire, looked out and saw a great light just oposit my stateroom on the quarter [deck]."

> Mr. H. went out and discovered that a bail of cotton was on fire. The carpenter coming out of the wheel house where he had been mending the wheel droped the candle on some loose cotton and it was all soon in a light blaze. They soon called the hands together got water and outed it, it was very fortunate for us that it was so soon discovered, for our Negroes were on top of the cotton and if it had of spread, much farther it could not of been stoped.

Now the travelers had a few days to spend in New Orleans. Probably they had to wait for a boat to transport them to Galveston or Matagorda. The *Galveston News* (of September 14, 1848) had advertisements of ships and packets connecting with larger lines with routes from Galveston to New York. Family letters do not indicate how the group reached Caney from New Orleans. While they waited, Ariella wrote to her mother Mr. Hawkins was so busy that she seldom saw him except at mealtimes, but she was enthralled with the delights of New Orleans.

They had met a gentleman and his wife and children who were also going to Texas, and Ariella and the gentleman's wife apparently struck up a friendship: "We went out together yesterday to look at the city. I differ from Mr. H. in opinion about the looks of the city it is one of the prettiest places I have seen yet Philadelphia not excepted. I never saw such a show of pretty goods and pretty things in my life. Oh me if I just had plenty of money what is it I could not buy here."

One of the things they needed to buy in New Orleans was furniture for the house they would occupy on Caney. J. B. Hawkins described the house "bought of Quick" as plain and needing to be patched up for their occupancy. Ariella's letter confirmed that the house they expected to move into on Caney was modest, although she had not yet seen it. When she went about New Orleans acquiring the necessary things to set up housekeeping, she says she selected "the plainest sort to correspond with the house." This structure in later years was called the plantation house.

You would like it in this section of the country, Ariella told her mother. Writing in November, she said the weather was warm and felt more like the month of June. She described their meals at a hotel where presumably they were staying: "We have garden peas, lettuce, butter beans and other vegetables every day for dinner, it looks very strange to see the people with summer clothes on." Ariella's daughter Sallie asked her mother to tell her warmth-loving grandmother (the child's namesake) she would very much like this country because there had been no days cold enough to sit by a fire.

Two years after his brother J. B. Hawkins moved to Texas, Frank Hawkins in 1848 brought his family from North Carolina to settle in Mississippi, where he had set up a plantation on an earlier trip. They traveled overland to Middleton in Carroll County, Mississippi, in a company of fellow travelers probably making a train of wagons and carriages, and averaging twenty-four miles per day.

Letters written by Frank and his wife Ann describe romantic scenery and a dramatic encounter. They had camped one evening within seven miles of Abingdon, at the southwest border of Virginia. Ann said the country was beautiful—"Cows and hogs are feeding all day in the greatest lots of clover and grass I ever saw. The people have so much butter." And they came upon a curious sight: "We passed a dwelling yesterday where there were four dwarfs, 2 girls and 2 boys. Mr. H. became acquainted with the oldest man 38 years old and a head lower than John, inquiring about corn and fodder. He conversed so well. He asked him how old he was and carried him on his horse to show us he is a curiosity, his face getting wrinkled, quite intelligent, seemed to be."[3]

Another camp site on top of the Blue Ridge Mountains she found sublime: "We came up the mountains last Tuesday, camped on top of the Blue Ridge that night. The scenery was very romantic and beautiful. We have not had a drop of rain since we left home. The children find a good many chestnuts."

Ann also gives a clue about why they left home. In this letter she refers to what she considers the "Christian duty" they owed to their slaves. With this expression, she hints at the reason for leaving: "Dear Mama [her husband's mother] though we are far from you and Papa we never can be removed from you and if it had not been for the Christian duty we owe to our slaves we never would have left you and I still entertain the same desire to return if we can arrange it in a proper way as to do so."[4]

She must have meant that the care they owed their slaves required having land enough to make farming profitable and their care affordable. Because the family had a workforce, they needed land enough for their slaves to work, and they were on their way to find it.

On the way to Mississippi Frank Hawkins wrote to his father as often as he could find a store or post office to mail the letter. On November 1, 1848, he wrote:

> Dear Papa:
>
> I again write you a few lines to let you know how we are getting [along]. Mrs. Read [his mother-in-law] has recovered from her cold and is well and stands the trip much better than I expected. We are now all well. My horses look as well as they did when I left though we have drove hard. This is the 28th day we have been upon the road and if nothing prevents in 9 or 10 days I think we shall be at our place of destination which will be a quick trip. Brother James and myself were 42 days upon the road when we went out 4 years ago but I am traveling a different route, nearer and much better than the one we went. I am now in 4 miles of the Tennessee River and shall stop early in the morning and mail this at the Post Office. We have camped out every night but I feel as much at home at camp as in a good house. My expenses have been smaller than I expected. . . . The cotton crop in this part of Alabama is not good. I learn in Mississippi it is better. My wife and Mrs. Read join me in love to you and Mama and all the family.
>
> I remain your affectionate son,
> Frank Hawkins

In a postscript Frank tells of a slave named Randal belonging to his brother, Dr. Alexander (Sandy) Hawkins. Randal was now cheerful but was "fit for nothing unless he is constantly watched" and was a man of "bad temper" Frank had not been aware of before they left: "I will sell him for as much as I can." Evidently Sandy Hawkins had placed Randal with Frank in order to sell him, and Frank later reported that he had sold Randal for $740. He "will have a good home. Mr. Williams who purchased him is a fine man."[5]

By November 14, 1848, when the Frank Hawkins family arrived at the plantation in Mississippi that Frank had previously set up, he was pleased with the crops and the management of the overseer he had hired for the place. "My Negroes are all well and look well and are well pleased. They have what they can eat and I find them well clothed. Mrs. Henry [wife of his overseer] is a very smart woman. She had the little Negroes brought in last night for me to see them and they look well." His overseer had brought in a large crop of corn, about 3,000 bushels or 600 barrels "for the force" and had twenty-four fine hogs to kill that Frank thought would average 175 or 200 pounds. Frank was well pleased with his overseer, whom a Mr. Purnell had hired for him.

Frank Hawkins wanted to sell his North Carolina holdings and use the proceeds to acquire more land adjoining what he already had in Mississippi: "I should like to have enough and a little to spare to work my force." Currently, then, he had more hands than he had land for them to work and would need to hire out some of them. He was motivated to buy more land by the excess number of slaves he had.

On November 29, 1848, a court case against Frank Sr. and J. B. Hawkins was decided. It hinged on a provision in the constitution of the State of Mississippi that prohibited anyone from bringing Negroes into the state for sale or for hire unless it was the intention of that person to be a resident of Mississippi. "We proved," Frank asserted, "that it was brother James and my intention to move here at the time. We brought out the Negroes in 1844."[6]

Frank described the facts of the lawsuit to his father. A Mr. Grider, who had hired their slaves for an agreed-on fee, was now refusing to pay their hire, and he based his refusal on this provision in the Mississippi constitution. The disputed amount was a total of $485. Four hundred dollars of that amount was owed to James B. Hawkins and the remainder was owed to Frank. "This is the only case where any refused to pay. Mr. Grider had hired two of James's blacksmiths and "one of my women" in 1845. The law was repealed, said Frank, in 1846. "This is the first case of the sort ever tried in this county and it excited a great deal of indignation against Grider." Frank noted that Judge Miller was "in every sense an honest man and a gentleman" and "knew our family by character and charged the jury all the way in our favor."[7]

Happy family news brought Frank Hawkins's letter to his father to a close. Two of the younger Hawkins brothers, "Dr. Sandy" (Dr. Alexander Boyd Hawkins) and Phil (Philemon Benjamin Hawkins) were soon to be married. Like all the Hawkinses who had left home, Frank imagined the pleasure of celebrating these engagements among his kin. "If Mama gives a blow out and cotton was 10 cents, my wife and self would go to it."[8]

Chapter 3

Starting the Caney Sugar Plantation

In Texas J. B. Hawkins went into partnership with another brother, John Davis Hawkins Jr., and on October 17, 1846, these two brothers signed a written agreement to buy and cultivate land on Caney Creek. While 1846 is the date of their written agreement, the two brothers had selected their land earlier in 1845. The agreement does not specify that their plantation is to be for sugar cane exclusively; it speaks only of the cultivation of crops in general. This written and signed partnership agreement was witnessed by still another brother, Dr. William J. Hawkins. Following is a summary of the 1846 partnership agreement that marks the beginning of the North Carolina Hawkins family's farming and (in later years) ranching business in Texas.[1]

James B. Hawkins and John D. Hawkins Jr. agreed to purchase land in Matagorda County for cultivating crops along the Caney Creek, sometimes called the Caney River. From Thomas Williams they bought about fifteen hundred acres at a price of three dollars per acre, payable in installments and without interest. The Thomas Williams League is located just north of Sargent on both sides of Caney Creek at the place that was to include the acres of the Hawkins sugar plantation and the crossroads village later called Hawkinsville.

The partners also purchased an adjoining tract of four hundred acres from Abram Sheppard at a price of five dollars per acre. And they purchased two hundred acres adjoining the Sheppard acreage from "the wife of Jacob Quick." These included the Quicks' 120 head of cattle and 80 hogs, boosting the per acre price to eight dollars. The total cost for these several purchases of land and, in the case of the Quicks, some livestock, was $8,100, as stated in the agreement. The partnership was fifty-fifty. The two brothers intended to cultivate their land for five years, each contributing seventeen hands to work and each to pay for provisions, mules, horses, etc., for the first year. Thereafter, people and stock "must be sustained from the proceeds of the plantation."

The partnership agreement made allowance for expenditures already made by James B. Hawkins and requiring reimbursement. The two brothers itemized the expenses already incurred for moving hands to Texas and setting up a plantation:

$177.75 for people going to Texas
37.50 for tools
73.50 for mules and oxen
16.50 for shoes for People after getting there
21.00 for axes
25.00 for iron
500.00 for the hire of Leo Alston's hands for this year 1846 [Ariella's brother]
200.00 for the overseer for 1846
75.00 for the rent of Hinton Curtis land for 1846
100.00 for cotton gin
25.00 for horse mill

As it turned out, John D. Hawkins Jr. was to a large extent an absentee partner and investor, although he did make several trips to the plantation in Texas, staying for many weeks at a time; his wife Ann accompanied him at least once. But the details of setting up the plantation on Caney were attended to by James B. Hawkins, who was clearly the man in charge.

By the time J. B. Hawkins brought Ariella and their children to Texas in 1846, he had made several preparatory trips to buy land and hire the overseer McNeel. Probably McNeel was placed in charge as a temporary overseer while J. B. Hawkins himself was in transit. McNeel may have been a kinsman of a neighbor plantation owner named Pleasant McNeel, who had gone into the sugar business earlier and was in a position to give advice to a newcomer.

Choosing McNeel to do the work of starting up the plantation on Caney turned out to be an absolute disaster. He did not attend to any of the tasks he was assigned, and when J. B. Hawkins and his family arrived on Caney after their riverboat trip down the Mississippi, both Ariella and her husband were shocked at what they saw. J. B. Hawkins erupted with indignation in a letter to his father:

> When I got here I found that the man I employed last year to attend to my business (McNeel) had done nothing at all last year. He made no cotton and he did not have a cabin up. I can't conceive what he was about all the year. He cleared no land after I left. I have but a poor prospect for a crop this year. I am clearing, building. Also I have planted five acres of land in sugar cane, I expect to put in about sixty acres in cotton, clear land and make plenty of corn and potatoes and try to raise some hogs. I shant be able to kill more than ten hogs this winter. They pretty well destroyed our stock last year. The bear killed nearly all the hogs and they let the cattle get so wild it is a very hard matter for us to get a beef. I had sooner undertake to kill ten deer than one beef, but I will alter things very much from what they are after awhile. I found a good many of my Negroes sick when I got here and they continued sick until I could get them in good houses. They are very much crowded at this time but I will continue to build until I get house room for them. They are all well at present.[2]

The following day, January 16, 1847, writing to his brother William, he was still hot with anger:

> I think the man I had as an overseer last year made the poorest out of it I ever saw. He is not worth the powder & shot it would take to kill him as an overseer. When I got here he did not have the first cabin put up. I had to have the old house bought of Quick covered & patched up so we could have a place to go in. It answers very well for the time we shall stay in it. I shall take my family to Matagorda the last of April where I expect to keep them to educate my children. I have been very busy building Negro houses and a house for my overseer. I have not got through with them yet. . . . I found a good many of my Negroes sick when I got here with chills. I think it proceeded from exposure camping out pretty well all the year. I shall have them very snug in a short time.[3]

And Ariella, too, writing to her mother on February 27, 1847, was outraged by this neglect: "We found nothing in the world done when we got here, neither a foot of land cleared nor a house built. But Mr. H. has now ten excellent cabins put up with plank floors, had to saw the planks by hand. A smoke house and blacksmith shop [word illegible] cisterns made, and cleared a hundred acres of land. He has planted some sugar cane and thinks by fall twelve months he can get to making sugar."[4]

An incident on a nearby plantation, where not just neglect but an outright atrocity occurred, was caused by "poor management," in J.B.'s opinion. Writing to his brother William on January 15, 1847, he described how a slave on the plantation of R. Blunt murdered the overseer, Mr. Pascal. One of the slaves on the place "knocked him in the head with a grubbing hoe and buried [him] in the field and plowed over him. After killing Pascal, the Negro shot himself through the hand and came to Matagorda [where] his master was living and told him the overseer had shot him." Then the man took one of his master's horses and in two or three days the horse came back with the saddle on. In J. B. Hawkins's interpretation, this murder was caused by the fact that the workforce on this plantation had been "badly managed" from the start. They had been hurriedly brought to Texas from Mississippi to prevent their being sold for debt. In any case, the dead overseer was dug up and decently buried. J. B. Hawkins offered the opinion that the "Negroes responsible will certainly be hung." We do not learn the end of the story or any of the precipitating motives for the killing.[5]

As J. B. Hawkins was building his cabins for his hands, he also had to plant and grow enough food for them. To raise corn and hogs was a first priority. Corn kept well and could be ground for grits or corn meal. Hogs were an efficient source of meat because a sow produces as many as a dozen shoats, and pork can be put into casings and smoked to preserve it. One of the most important structures to be built first on the plantation was a smokehouse. Milk cows, chickens, fruit trees, and a large vegetable garden helped fill out the food requirements. Fish, deer, and ducks were also plentiful along Caney Creek.

The Caney plantation's commercial crops—sugar, molasses, and cotton—had to be planted, processed, and marketed. The plantation was a whole complex of activities and buildings: dwellings, a smokehouse, cisterns, a blacksmith shop, barns, a gin, saw mill, brick kiln, and by 1850 a brick sugar mill for boiling down the juice of sugar cane to molasses and brown sugar. On the place were carpenters, brick masons, and other skilled craftsmen and handlers of horses and mules.

Work life and home life were interwoven on the plantation for both black families and white families. The work on the plantation and whether it succeeded or failed were as fully visible to the wife and children of James B. Hawkins as to him. Both Ariella and the children often wrote to their North Carolina relatives about the number of acres that "Pa" had now planted and how many hogsheads of sugar or molasses he expected to make. All who lived in the plantation community—blacks as well as whites—had a stake in the success of the enterprise, and all had to rise above hardship, crop failure, storms, illness, and the birth and death of children.

The town of Matagorda played a significant role in their lives almost as soon as the Hawkins family arrived on Caney Creek. Matagorda was a small but thriving port where boat traffic coursed back and forth from Galveston to Indianola and made connection with oceangoing vessels of more distant destinations. The sugar business required J.B. to go there routinely. It was a place for families to attend church and enjoy neighborly visits and for older children to attend school and social gatherings. To planter families, escape from the swampy air of the Caney Creek region into the healthier open air of the Matagorda Bay front was a prime attraction of Matagorda. As early as January 15, 1847, J. B. Hawkins was making plans to take his family to Matagorda that summer and have them stay for several months, as they would do year after year. He wrote to his brother William: "I shall take my family to Mattagorda the last of April where I expect to keep them to educate my children." He had rented a house there for ten dollars per month.[6]

The Christmas season of 1847 brought surprising news from Matagorda of great significance to the later Hawkins Ranch family. Ariella described the mid-December event for her mother:

> I received your welcome letter last mail and would have answered it by the same but was prevented by the birth of a fine boy, he was born last Friday night, week ago [December 17, 1847]. I was entirely alone. Mr. Hawkins being on the plantation, he did not hear of it until he got to town [Matagorda]. It was then four days later. . . . I never did better—we, that is babe and myself are doing remarkably well. Mr. Hawkins has named him Frank after his brother. I have been up and walking about for a day or two. Mrs. Stuart and Mrs. Kelsey two ladies of this place have been very kind to me. They came and staid with me until Mr. Hawkins got home, and they come up every day to see me.[7]

Ariella's observation about doing remarkably well reflects her usual optimism. Probably she was not "entirely" alone in undergoing the birth of the new baby. Her husband was not with her, but there would have been servants in the house to help with the new baby's delivery and with her other children.

Frank Hawkins, the son born to her on December 17, 1847, at Matagorda was the first of her children born in Texas. He would be the father of the five—Harry, Meta, Janie, Lizzie, and Elmore—to be introduced in later chapters. Certainly the birth of the new baby was paramount for Ariella at Christmas time, but in the same letter she offers her mother a glimpse of the way the town of Matagorda celebrated Christmas in 1847: "Yesterday was Christmas. The people make a great to do here on Christmas. The whole town seems to be in a complete swirl. They kept up a continual show of firing guns from midnight Friday until midnight last night. They gave a large party to the little girls Friday night. I let Sallie go. She was perfectly delighted."[8] Frank's eldest sister Sallie was ten years old; her mother sometimes spelled her name Sally.

Ariella had invited her younger brother Edgar to spend Christmas with them in 1847, but unaccountably he had not come. The lack of response from Edgar foretold trouble for him. A depressed young man, he was then living in Galveston after military service in the war with Mexico. He was trying to get into business there but failing in the attempt. His plight, described in a later chapter, reveals the way slaveholding planters thought about the economics of slaveholding.

Of enormous interest to J. B. Hawkins and of equal interest to his wife and children was the construction of his sugar house. It was the main news item in letters to North Carolina and the subject of a painting by his daughter Sallie. Like other new sugar growers along Caney Creek, J. B. Hawkins set about building a sugar mill made of bricks, and the bricks themselves had to be made and fired in a brick kiln. A cousin named George Nuttal who helped out with cattle branding also served as a "kiln smith." The bricks were made by digging out the local clay and pressing it into rectangular brick molds—three brick-sized hollow rectangles, set end to end, in a three-foot-long wooden rack. The wooden racks were expertly crafted with mortise and tenon joining. Two narrow slots, running along the bottom of the wooden racks, served as sluice openings to drain each of the three wet clay bricks. The molds had to be made of sawed lumber, and the bricks, once molded and sun-dried, had then to be fired in a kiln, which itself had to be made of similarly produced bricks. J. B. Hawkins calculated the number of bricks needed to build his sugar house and based his estimate on the experience of other sugar growers he visited between the San Bernard and Brazos rivers. He made a systematic investigation of what other sugar cane growers were doing, was guided by their experience, and gave a report to his father and his brother William.

A planter named Bob Miller had 500 acres planted and expected to make 1,000 hogsheads of sugar. His sugar house took 700,000 bricks to make. And Pleasant McNeel and his brother had 300 acres in cane and expected to make 600 hogsheads of sugar. McNeel's smaller sugar house took 500,000 to 600,000 bricks. "Their whole expense including building, sugar engines, and 2 sets of kettles, everything to make 18 or 20 hogsheads of sugar a day will cost them ten thousand dollars. They will more than pay for it the first crop." Clearly J. B. Hawkins was enthusiastic about the sugar business and optimistic about his prospects in two years' time, 1849: "I will have cane enough to employ forty hands to take off the cane. I shall plant this year . . . over 300 acres, . . . so the fall after, say I make 2 hogsheads to the acre . . . I will make 600 hogsheads of sugar." He described himself to his brother William as "the busiest fellow you ever saw."[9]

J. B. Hawkins began growing sugar cane before he had completed his sugar mill, so he must simply have sold the cane to other producers or possibly contracted with them to boil it down for him. In 1848 he started making handmade brick for the construction of his own sugar mill. The construction began on the first of March in 1849, but J.B. was not optimistic about its completion that same year. He sent a detailed description to his father:

> We will have the main building for the engine kettles and coolers 45 feet wide and 160 feet long with a wing each side in a shape of a "T" of the same width and 80 feet long each. The main building to be 22 feet high and the balance to be 12 feet. We shall put land enough in cotton next year to make 150 bales if the year suits it. We made 6000 bushels of corn. The sugar is very good here this year, several farmers made twenty five hundred pounds of sugar to the acre. Brother John [John D. Hawkins Jr., his partner] has been with us some 4 or 5 weeks. He speaks of leaving for North Carolina the first of February, but you can tell Sister Ann [John D. Hawkins's wife] not to look for him until the first of April as I shall try to keep him here as long as I can. We are all in fine health, brother John and my wife join me in love to you, Mama and the family.
>
> I remain your affectionate son,
> James B. Hawkins[10]

By June 12, 1849, J. B. Hawkins is trumpeting to his brother-in-law Major Arch Alston: "I am getting on finely with my sugar house. It will be the finest in Texas & the largest. Such a brick house you never did see. It will take nine hundred thousand bricks to build it."[11]

That fall Ariella reported to her mother that the outside of the sugar house was completed. "It is built of brick in the shape of a T. It is three hundred and fifty feet front, two hundred sixty back," she wrote. She wished her husband would be ready to grind his cane that winter because he "has cane enough [planted] to make two hundred hogsheads."[12]

The image of a plantation romanticized in literature and film and suggesting a lazy, self-indulgent, and trivial life is not the life depicted in the letters of these Texas pioneers. Historian Howard R. Lamar used the phrase "the entrepreneurial plantations of pre–Civil War Caney Creek" to describe these sugar plantations.[13] Certainly J. B. Hawkins viewed his plantation in just this way, as a business enterprise, for in letters he described himself as a very "busy fellow" who directed his action toward profit. He habitually referred to all his activities as "attending to my business."

Chapter 4

ARIELLA AND PLANTATION FAMILY LIFE

Ariella had great confidence that her husband would succeed. She and the children knew all the details of her husband's projections for the planting, harvesting, and marketing of his crops. Her own day-to-day responsibilities included the care of her children and managing the household, the servants under her supervision, and the vegetable garden and poultry. She was as busy as her husband with her eight children, her 102 chickens (her actual count), and a garden that she told her mother provided peas, lettuce, radishes, pumpkins, Irish potatoes, tomatoes, and cabbages "in abundance." And she hoped that her mother would send her more seed, which she said was hard to get. She also noted she had sewn three suits each for her two little boys, James and John.

At first Ariella plainly missed her relatives and felt shut off from the socializing so available to her back in North Carolina. Her reports of this isolation are factual and uncomplaining. "I never have anything to write about except my family," she said. At Caney she had neighbors, but only one was near enough to visit in a day and return the same night. She explained that the close neighbor was the family of Colonel Jones, who happened to be the "brother of the girl I went to school with in Salem . . . Mrs. Jones has been to see me twice. She is very clever and kind and I like her very much."[1]

From time to time Ariella gave her mother a description of her children's activities on the Caney plantation. Sallie, who was ten, was knitting Virginia a pair of gloves "just now." Sallie also played with a pet deer "that one of the Negroes picked up in the woods." She named the deer Sallie "after you," Ariella told her mother. But then the child found out it was a buck, and so she was calling it Jimmy. Virginia (seven), James (five), and John (three) were "building a house at the door out of blocks to keep store in. They seem very busy and highly amused." Little Ella was a toddler one year old and could say a few words. "She is the smartest and sweetest thing you ever saw. She is at this time sitting on the floor eating beef heel and potato. She is so fat that when she is down she can hardly get up." Ella would not live past her sixth birthday.[2]

The Mrs. Kelsey who had come so often to visit Ariella at the time of Frank's birth in Matagorda had become a great friend of the family. On November 9, 1849, she was visiting them at Caney. Ariella found her "a most excellent little woman," helpful in every way in keeping house, sewing, tending the sick, and assisting Ariella with any number of her household chores. Ariella wrote that Mrs. Kelsey and Virginia were just then busy making a cake but that Ella and Frank, her youngest children, were "singing so loud in the passage that I can scarcely write."

> Sally is with her pony. Willis, James, and John are off to some sugar house or other. Our little boy Frank is one of the likeliest boys I ever saw. I dress him in short sacks, knee britches, long stockings, and shoes. Tell Sister Mary and Mzury if they could see him they would never brag about their boys again. He has been walking more than a year but can't speak a word yet, still makes more noise than all the rest put together, laughing and singing. . . . The Negroes are all well and send howdy to their kin folks. Becky and Emily say give their love to all white folks and Negroes too, tell the Black people howdy for me.

During leisure times, horseback riding was a special pleasure for the children and the whole family.

> The children enjoy themselves very much riding horseback. They go first to one sugar house and then to another. They always come back loaded with sugar and molasses: James and John say they love the sugar houses because they are so sweet. They are all first rate riders. Their Papa bought them all a mustang pony apiece the other day. They are not broke yet so they ride double. Sally and Virginia, James and John, Willis and Frank. Mr. Hawkins takes Ella, and I ride by myself on one of the prettiest horses you ever saw. We call him *Red Bird* from his color, he is very much like Brother Arch's *Fashion* only handsomer. Sallie says I must tell you we make a great show when we travel. We very often go twenty and thirty miles a day and make nothing of going seven and eight miles and back after dinner. I still hope you will come out this winter, I will not suffer myself to think to the contrary for a minute. I am as great a hand as ever to hope for the best if everything does go for the worst, as we poor short-sighted mortals are too apt to think.[3]

The overseer had promised Sallie that he would ride her pony for her to make the animal gentle enough for her to ride. "She tells him he must make haste and break it and make it very gentle, for she is a going to give it to Grandmama to ride when she gets here, it is a natural pacer," Ariella wrote. The children "all think you will be here this winter and I cannot bear to tell them any better."[4]

Ariella had many hands to help her with her children. Family letters suggest that a "mammy" was assigned to each of the smaller children, who needed to be bathed and dressed and constantly watched. Older children rambled freely through the woods gathering flowers or rode off on horseback to visit other plantations. By 1850 little Ella was walking and able to hold her own in argument with her brothers and sisters. The children were always trying to hush up "Punk," their nickname for little Ella, "telling her she doesn't even know anything about Grandma, Uncle Arch, Aunt Mzury, Uncle Charles and all of um, an you keep such a talking up we can't talk none. El will answer him very quickly, 'I sank you sir I do know my Ganmar and I will talk. I [got just as] much right to talk bout my Ganmar as you ain't I Mar.' She always appeals to me to know about her rights."[5]

It is revealing that the children judged their standing by how well they knew their North Carolina relatives. The whole family, living in Texas far from Hawkins and Alston relatives, placed great value on keeping the family links well fastened. Ariella pleaded that her mother teach their names to new babies born into the North Carolina family so they "will learn to know us." The Texas family returned to North Carolina almost every year, and it is surprising but true that when the time came for formal education beyond what Matagorda could offer, several of the Hawkins children were sent to schools in North Carolina.

In 1847 as she settled in, Ariella genuinely seemed to like her Texas home on Caney Creek and worked energetically for the next two years to make its landscape beautiful. Her husband began to call the plantation "Magnolia" because of the many magnolias in the yard, but no one else seemed to use the name. Sallie said their yard was full of a variety of trees—"magnolia, orange, wild peach, elm, sycamore, prickly ash, oak, yaupon." And she said her mother had set out rose bushes and planted a great many mimosa seeds. "She intends having the yard laid off and planting every kind of shrubbery in it."[6]

By April 11, 1847, Ariella gave Texas and their lives on Caney a sweeping affirmation: "This is the greatest country in the world. We have to undergo some inconveniences in moving to as new an environment as this is, at first, but what is that when we see we will be doubly and triply paid for them in a few years. This part of the country is settling up very fast. There has been three or four families to move in and settle since we got here."[7] And she sent her brothers Charles and Leonides an added teasing word: "Tell them to come to Texas if they want to get rich . . . Mr. Hawkins says he expects to get so rich that he will be able to drive sixteen horses in hand when he pays a visit to North Carolina."

Brother Arch, Ariella's sister's husband, should have some of their fine beef, she said. "I get more tallow from two beeves here than I ever got in North Carolina." The winters, she explained, "are mild and so we have to wait for a norther to kill meat and then only a little at a time . . . we killed seven hogs at one time and I saved nearly a whiskey barrel full of lard. You may judge the size of them by that."[8]

Cattle were plentiful but wild and, in the absence of pasture fencing, were difficult to brand and mark. J. B. Hawkins had bought some cattle that came with the land he purchased; in addition, in appreciation of a favor, his elderly neighbor Hinton Curtis had willed him land with cattle on it. From the inherited Hinton Curtis herd, J.B. and George Nuttal were able to mark and brand forty calves, which were fewer than half the calf crop, according to Ariella.[9]

Apparently Ariella needed to emphasize the attractiveness of this new country, as her family in North Carolina had obviously been a little skeptical. She often urged her mother to visit, assuring her that the journey need not require travel on the Gulf of Mexico. Cousin William Harris had returned to Mississippi by land, "so you see we can get to and from Texas without crossing the Gulf . . . the Negroes are all well and send howdy to their kin folks. Becky says tell ole Mistress and Miss Missouri howdy." Before Ariella closed this letter, written on April 11, 1847, she waited for Sallie and Virginia to return from the woods in order to send along with the letter some "lilack and snow drops," which she said grew all over the woods.[10]

In 1849 the family went as usual to Matagorda in the spring and stayed until November. During that time the children attended school. As winter approached, Ariella felt she needed to take the children out of school and return to the plantation, because the house in Matagorda was not adequate for winter weather; it was "so very open." This experience with an inadequate house at the Matagorda Bay shore might have prompted J.B. and Ariella to discuss building a house instead of continuing to rent in Matagorda. But they were not quite ready to undertake that. At this time, while the sugar house was nearing completion, processing sugar cane was not in full production. "Mr. Hawkins says he can't think of putting up a better [dwelling] until he makes a crop or two of sugar. He promises me great things when he makes sugar." Here is a hint of their interest in one day having a larger house for their family, located where it could catch the healthful bay breeze. It was an early wish leading to the building of the Hawkins Ranch House, which J. B., Ariella, and their children would call their "Lake House" because it would be built where Lake Austin, in the shape of a cumulus cloud, spread inland from the Texas coast.[11]

Sallie reported to her grandmother that they had intended spending the summer of 1849 on Caney Creek but that when her little brother Frank, aged two, was taken sick, her father said they must all go to Matagorda. "The wagon will go down on Monday and we go on Wednesday." Moving toward the coastal breeze was the first remedy anyone thought of when illness struck.[12]

The illnesses from which they sought relief by the coastal breeze were called "chills and fevers," symptoms that describe malaria. The word *malaria* was never used and not then known, nor was the cause of the disease, the bite of parasite-carrying mosquitoes. People on the plantations along Caney associated chills and fevers and cholera with polluted air blowing across swampy ground. The remedy was to move away from low places to the open air. The three medicines used in some combination to treat almost any illness were quinine (which was probably effective), calomel (a cathartic and fungicide), and laudanum (an opiate). When J. B. Hawkins himself was stricken with chills and fever, Ariella's mother sent her a recipe for a tonic with French brandy as its base. Ariella's method of treating cholera included doses of calomel, red pepper, and camphor (see chapter 5).

By April 28, 1850, the all-important sugar house was finished. Sallie, a budding artist, painted a picture of the sugar house to send to her grandmother. A copy of Sallie's painting was done in the 1940s in Bay City, Texas, by Georgia Mason Huston and is now the only representation of the building apart from the verbal description and measurements given by J. B. and Ariella. J. B. Hawkins was in Matagorda and had time to write his mother-in-law a letter, to which his wife added a postscript.

Matagorda, Texas
April 28th 1850
Dear Madam
. . .
I brought my family to town [Matagorda] on Wednesday last to spend the summer. We are all well and look well. I think I can say with the utmost sincerity that I have the loveliest wife and eight of the loveliest children that can be wanted in any county and I can't help from being proud of them.

My wife brought 4 Negroes & 2 children here to wait on her, Horrace, a good man servant: Emily & child, Sylvy & child and Becky. I shall go back to the plantation after I get them well fixed here, to stay most of my time, our crops look well. The Negroes all enjoy good health.

We had a most excellent sermon this morning from our Episcopal minister Mr. Denison, he is much esteemed here by all. We received the box you sent us a short time ago and we all send much obliged to you for the kind presents it contained.
. . .
My wife is taking a siesta, I will arouse her & let her send her own love to all. J.B.H.

[Ariella's postscript]
My Beloved Mother
Mr. Hawkins has left me room to add a few lines. We moved to town a few days ago and have fixed ourselves very comfortably—provided we have no more cold weather. I forgot whether I ever told you that we have had the house partitioned off and made five rooms where we had two before. I have a nice little front room for you I sincerely hope before another summer you will be here to occupy it. Accept, my dear mother, our thanks for the presents you sent us. We are all very pleased of them and they were very things we needed most. Becky sends howdy and a thousand thanks for her presents. I wish you could see our youngest child [Charles Edgar, called Edgar]. He is but three months old and can pat a cake, laugh and cry so as to be heard all over the house. The children have all improved a good deal. They will start school Monday. The Negroes were all well when we left the plantation. Sallie drew and painted the sugar house last week. She sends it to you and says you must excuse the painting as it was her first attempt. I planted the plum and coffee seed you sent. Please get me some of all sorts of fruit seed and bring with you next fall. Excuse this bad written postscript. Give my love to all my brothers and sisters. Sallie and the children send love to you and their uncles and aunts. Kiss the children for me.

I remain your affectionate daughter, A Hawkins[13]

By December 1850 sugar production at the plantation was fully under way, but the schedule was running far too slowly, and J. B. Hawkins was in danger of losing his cane in the field to spoilage. They attributed the slow pace to a lazy sugar maker. Ariella gave her mother a full report of the trouble. Every three days, the "sugar maker would get sick or feign sickness." She was afraid this man might "cause Mr. Hawkins the loss of sixty or seventy hhd of sugar, for we had a very severe freeze lst week, the severest ever known in this county, and Mr. Hawkins thinks he will not be able to gather all his cane before it spoils."

But J. B. quickly replaced the lazy sugar boiler, and the production was soon running at full tilt with a more "industrious and excellent sugar maker in his place," making every effort to "save the cane before it sours," she wrote.

Showing that she knew all the details of her husband's work, Ariella told her mother that "he has a hundred and forty-eight hhd up and bout twenty in the coolers. They are grinding now making a hundred hhd in twenty four hours." She said her husband might run "both sets of kettles tomorrow" to speed the process. Especially in this harvest season, James B. Hawkins was working long hours to get his crop gathered. He was at the sugar house as she wrote. "He keeps the first watch and has to stay up until twelve."[14]

Chapter 5

THE CASE OF EDGAR AND WAYS OF THOUGHT IN SLAVERY TIMES

Although Spain and Mexico had opposed slavery, when Stephen F. Austin was colonizing Texas and about twenty-five years later when the Hawkins family came in 1846, the state was seeking settlers who could farm profitably and contribute to a productive, stable citizenry. Successful planters were those who came with their own labor supply—their slaves. Texas had a surplus of land but a scarcity of labor. Thus families from the South who came with a slave labor force were welcome, and the slave-holding families found the cheap land in Texas a boon. Growing crops was profitable provided there were enough laborers to farm on a larger than subsistence scale. One contemporary observer notes that "very few even poor men consent to be hired, preferring to work their own lands," a disposition that reduced the number of acres worked and made them less profitable than a more extensive slave-worked plantation. For this reason, almost no one at the time could think of a profitable way to cultivate a large plantation except by slave labor.[1]

In the Texas of the 1840s and 1850s, cultivating crops was the main road to economic success, and almost the only road. If the land holdings and slave holdings were large, the planter could thrive. There was little or no commerce or industry in Texas to offer an alternative opportunity. "Storekeepers made a living in small towns across the state and a few large factories thrived in Galveston and Houston, but commerce constituted a relatively minor sector of the economy." In 1850 and 1860 less than 5 percent of the state's household heads reported occupations in commerce, and there was virtually no manufacturing or industrial employment beyond local agricultural machine shops. The planters who thrived by combining large land holdings with large slave holdings had no motive to risk investing in commerce or industry. They were motivated, rather, to increase their number of slaves and, as their numbers increased, to increase the acres of farm land to be worked. In Matagorda County the slave population increased from 1,153 in 1850 to 2,369 in 1864.[2]

Ariella's young brother Edgar Alston came to Texas and wanted to get into commerce rather than farming. Given the scarcity of commercial occupations at the time, this choice all but doomed him to failure from the start. But the important added factor was that he had slaves. How many is not mentioned in correspondence, but owning them and hence having the responsibility and expense of their care further compromised his economic success.

Sometime during 1845, about the time that James B. Hawkins was preparing to settle on Caney Creek, Edgar left North Carolina to live in Galveston. He did not choose Texas because his sister and her husband intended settling there, nor was his presence in Galveston an incentive for their move to that region. Edgar's venture was a restless young man's impulse to leave home and explore a new territory. Like other young people from southern states, including some Alston cousins, Edgar had a desire to explore this new place, Texas.

Unlike J. B. Hawkins, Edgar did not come to Texas searching for land in order to cultivate crops; he had no interest at all in farming. Edgar wanted to go into "business," to find a clerkship with a merchant. The busy ports of Galveston and New Orleans seemed to be good prospects for him, but those prospects soon dimmed.

A letter of December 22, 1845, shows that Edgar was then living in Galveston. He had been in touch with his brother-in-law J. B. Hawkins and with John D. Hawkins Jr. as they passed through Galveston on their way to retrieve part of their workforce from Mississippi, a workforce J.B. had probably left there on the trip in 1844 with his brother Frank. Writing to his mother from Galveston, Edgar reported: "Mr. Hawkins and John D. left here this morning for New Orleans. Mr. Hawkins is going to Mississippi after his Negroes and is going to return as soon as possible to settle them on his farm. Brother [James B. Hawkins] and John got lands joining each other on Caine [Caney] Creek. It is said to be navigable above their plantation."[3]

Edgar allowed himself to be stymied by obstacles and tended toward self-pity. In a letter to his mother he whined, "This is the third letter I have written home, and I have not so much as received a line from you to let me know that you are well, nor neither have I received one from either of my brothers or sisters." Edgar would turn out to be an unsettled and dispirited young man who failed to find the stimulus he expected in venturing to Texas. Given the slave agricultural economy of the times, the fact that Edgar continued to own slaves without using them to farm was a major source of his economic troubles.

His unrest, gloom, and lack of employment became the subject of worried letters that passed among J. B. Hawkins and Ariella's elder brother Charles and also between Ariella and her mother. Even so, Edgar kept insisting to his mother that he was "better pleased with Galveston the longer I stay. . . . I find it a place full of life and enjoyment. I attended a splendid ball last Thursday night. There were a great many people there and all of them seemed to enjoy themselves very much and I also enjoyed myself as well as could be expected being a stranger and unacquainted with the people."[4]

Young men were taking a look at Texas just as it was coming into the Union to see what promise it might hold for them. Edgar reported to his mother that his cousin Thomas Alston and a friend named Brodie had been down to look over Texas, but he did not think either would settle there. "Tell Aunt Bettie that she may not make herself uneasy about Cousin Thom for I don't think that he could be hired to stay in Texas."[5]

Edgar's spirits were given a bolt of energy by national events when Congress declared war on Mexico on May 13, 1846. Edgar, aged nineteen and full of patriotic fervor, was keen to volunteer for the fight. This worrisome piece of news spread to the North Carolina family and to Ariella and J.B. at their Caney plantation. Mrs. Alston was, of course, alarmed. She was a worrier anyway, even about small things, as Ariella's many words of reassurance indicate, and Edgar was the baby of her family.

Ariella, characteristically optimistic, wrote to her mother that there was probably no need for worry: "I have been expecting to hear [the news of Edgar's enlistment] ever since War commenced. I do not feel at all uneasy about him, for Mr. Hawkins says and I think there will not be any need for his services. I see from the papers that General [Zachary] Taylor has more men than he has use for."[6]

The war with Mexico had come to a head after Texas won its independence from Mexico in 1836 without Mexico's recognition of that independence. When Texas joined the Union as a state, Mexico broke off relations with the United States but did not declare war. Quarrels about the boundary of the new state of Texas emerged. Was it to be the Nueces River, as Mexico claimed, or the Rio Grande? President Polk tried to negotiate the boundary dispute by offering to pay Mexico for the territory, but these attempts failed, and when General Zachary Taylor moved his troops beyond the Nueces into the disputed territory, he was attacked. In the end Mexico lost territories that became the states of California, Nevada, Utah, and Arizona as well as parts of Colorado and New Mexico. The war declared on May 13, 1846, ended on February 2, 1848, with the treaty signed in the village of Guadalupe Hidalgo, near Mexico City. By this treaty the United States acquired the territory that Mexico had refused to sell.

Edgar did serve in the war, and for the rest of her life his mother kept the letter written by Memucan Hunt quoting the commendation of his war service in Monterrey: "The celebrated Captain Walker who was Lieutenant Colonel of lst Regt Mounted Riflemen stated to me that 'Major Alston . . . was immediately under his eye in the battles of Monterey [sic] and that a braver cooler or more efficient soldier was not in those fights.'"[7]

Edgar's military service ended before the signing of the treaty, and he was still determined to live in Galveston. He was having a hard time getting a job as a clerk but intended to try both in Galveston and Houston. Eventually Edgar was able to report that he had "got into business" as a clerk for a gentleman who seemed to be a fine young man and who took pains to show Edgar the details of his trade. Edgar never stated what kind of clerkship he had obtained or what salary he was paid; as it turned out, he served an apprenticeship with no salary.

By February 1847, Edgar was clearly hard up. Ariella and J.B. asked him to come and stay with them for a time, and he accepted their hospitality both at Caney and in Matagorda. Ariella told her mother that "Ed is with us and will go with us to Matagorda to see if he likes it as well as Galveston. Ed has hired his Negroes out and intends to go into business in Galveston."[8]

Edgar's reduced circumstances were sadly revealed when he wrote to tell his mother that accepting the hospitality of his sister and brother-in-law would save him $265 for his room and board in Galveston. Mrs. Alston then came up with a new worry about Edgar. What if he got married? How would he support a wife and family? His reply groans out his gloomy misogyny: she need not worry about his getting married, he wrote, "for I can apprise you that I have never seen that woman that I have had the least conjugal love for and the more I see of the world and of woman, the less I care about sharing their troubles or pleasures." So much for the young ladies at the Galveston ball.[9]

Edgar was clearly being overtaken by depression. Edgar's situation was worrying enough for J. B. Hawkins to write Charles Alston, the elder brother of Ariella and Edgar. According to J.B., the main reason for Edgar's economic troubles was that he owned slaves but was not going to use them to farm. And if he was not going to farm, in the opinion of J.B., he certainly should sell them. With the proceeds, J.B. believed, Edgar could set himself up to trade in flour and such commodities in New Orleans and "could have doubled his money." J.B. said Edgar "writes very discouragingly and speaks of going to Mexico."[10]

Edgar was hedged in by limited choices imposed not only by his own temperament and work preferences but also by the economic constraints of the era of slavery. He could use his workforce to farm, but he rejected that option in favor of going into business, unsuccessful though he was in doing so. His other options were either to sell his slaves or to hire them out for a fee paid to himself.

At first Edgar's mother was a great impediment to his selling his slaves, and she had a great deal to say about the matter. Edgar's slaves had probably once worked on land in North Carolina that belonged to her family. She was concerned for their welfare and did not want them sold out of her family and running the risk of mistreatment by strangers. She absolutely and insistently forbade Edgar to sell them, as he reported to J. B. Hawkins, who had urged him to do so. Edgar pleaded to J. B. Hawkins that he must be a dutiful son and go along with her wishes.

The option of hiring out his slaves was a tactic he tried, but he was not able to make an arrangement steady enough to provide him a secure income. J. B. Hawkins explained to Charles that Edgar's workforce did not include skilled workmen like blacksmiths or carpenters or masons, who could be easily placed out at hire. In J.B.'s explanation to Charles, the cost to Edgar of the care of his slaves was more than the income derived from their sporadic hire. Edgar was stymied, and the results harmed both himself and his slaves, who were cast into purposeless idleness without their freedom and without even the dignity of work. Edgar's slaves, according to J. B. Hawkins, had been seen on the streets of Galveston unemployed, loitering, and drunk.[11]

To free his slaves—the option that presented itself so forcefully to abolitionists of this period—apparently never rose to the level of consciousness in Edgar or other slave holders. They could not imagine how farming could be conducted in the absence of a slave workforce or what possible employment the slaves could find in the absence of farming. For them, slavery and large-scale farming were inextricably joined.

What deafened the planters to the abolitionist plea to set slaves free—a plea given little or no actual voice in the region—was their economic assumptions. The planters regarded the workforce as a capital asset that could be freed only at the insupportable economic cost of ceasing to farm in the only way they knew how. They further believed that if their slave-dependent system of farming failed, the slaves would become rootless, homeless, and devoid of the care planters thought of themselves as providing. That the sudden liberty of slaves might trigger unrest or at least homelessness seemed to be in the mind of Gen. Gordon Granger even as he announced his Emancipation message of June 19, 1865, at Galveston. He then advised newly freed slaves "to remain quietly at their present homes and work for wages."[12]

Out of desperation, Edgar finally did sell his slaves, and he was severely reproached by his mother when she learned of the sale. On December 26, 1847, Ariella wrote that she had not heard from Edgar in some time. She hoped he would spend Christmas with her family on Caney, but he had not come and had not written. Loyal to her brother, Ariella interceded with her mother on his behalf. "I am sorry he displeased you so much by selling his Negroes. I think you were displeased without a cause for he wrote to you asking if you had any objection but never received an answer." Edgar had told J. B. Hawkins that his mother had forbidden the sale, but he had evidently told his sister that he had not received a reply from their mother as to whether she objected. Possibly Edgar did not tell his sister the full story: that he had sold his slaves because of his desperate financial need.[13]

Edgar died in Galveston in the fall of 1848. His obituary in the *Galveston News* was dated September 14, 1848. The piece praised the deceased in the elaborate language of the day without revealing the cause of Edgar's death or much else about his life. Clearly the obituary writer did not know Edgar and relied on ready-made language to commend him. Edgar was only twenty-one.

J. B. Hawkins went to Galveston to settle Edgar's affairs. On June 12, 1849, he wrote to Major Arch Alston that Edgar had had room and board privileges with a Mr. Bennet in Galveston but that Ed did not owe Mr. Bennet anything; nor was he owed anything by Mr. Bennet. "He was getting no salary at the time of his death," wrote J. B. Hawkins, revealing that Edgar had gotten nothing in the way of salary, except his board, for most of the time in Galveston. His slaves, when he had them in Galveston, made him no money but ran him into debt, J. B. Hawkins explained, adding that doctors' bills for Edgar's sick slaves were greater than income from the hire of the well ones. "He managed much worse than I thought he did! . . . He never would consult with me much about his business. He . . . was never satisfied unless he was attending to some business whether he got paid for it or not." He did not "have a cent of money at his death." Nor did he owe anyone except J. B. Hawkins, who held his note for nearly four hundred dollars advanced to him. J.B. wrote that he would "do the best for all hands in settling up his concerns here."[14]

The case of Edgar illustrates how planters of the day viewed their slave labor force and calculated the economics of farming. If the planter had land or could acquire it cheaply, then the force was an asset. But absent land enough to work or absent the intention to farm, such a force turned into an owner's liability because of the continuing cost of upkeep in housing, food, clothing, and medical care. For Edgar the cost had been ruinous, both to himself and to the slaves depending on him.

As a whole, these family letters illustrate three ways of thinking and speaking about slaves: commercial, personal, and racial. The commercial way was more usual with men than women. In their letters men used the term *workforce* when thinking about the business of farming. They spoke of the number of bushels of corn they had harvested to sustain "my workforce." Sometimes they considered the need to acquire more land in view of the growing size of the workforce.

Slaves were often inherited from the estates of deceased family members. Slaves could be hired out at a certain rate to a planter or to the county for road repair. They could be sold to pay a debt. James B. Hawkins and his partner-brother John D. Hawkins Jr. borrowed the slaves owned by Ariella's brother Leonidas and acknowledged in writing that they owed him five hundred dollars in rent. Buying and selling slaves was not a central aspect of business for the owner of a plantation but was certainly incidental to its operation. As J. B. Hawkins began the operation of his plantation, he wrote to tell his brother William that "brick masons, good carpenters, and good blacksmiths will be in great demand. So soon as farmers make a crop, Negro field hands will sell well."[15]

On January 15, 1847, J. B. Hawkins wrote to his father:

> I have sold your two Negroes Lucy and Osborne, Lucy to the overseer, Mr. Wright for $600 and Osborne to Mrs. Quick for $700, so I owe you $300. I have paid Mrs. Quick all we owe her so brother John will owe me $300 on that score. You will please tell him to pay you $300 and I will settle with him for it and pay him the interest on it up [to] the first of January, 1848.
>
> . . .
>
> I have sold Jim, the boy John bought of Mr. Andrews and kept William's boy George in his place. Jim is a very trifling boy. I sold him to Mrs. Herbert for $750. I could have gotten $50 more for George but I would not give him for two of Jim. I understand Jim has run away and is out at this time. I sold him on credit, the note bearing eight percent interest.[16]

In another transaction John D. Hawkins Jr., living in North Carolina and finding himself in debt to Ariella's brother Charles, sent him a letter proposing to pay the debt by selling his slave Nancy and "her child or children." He enclosed in the letter to Charles an order to sell Nancy and her offspring. Because Nancy lived and worked at the Caney plantation the order to sell is addressed to J. B. Hawkins, who had been apprised of the transaction and was asked to execute it. The order was enclosed in John D.'s letter to Charles almost as if it were a check, a kind of negotiable instrument. The order required J. B. Hawkins to have Nancy and her child or children "valued" and sold, and then to send the proceeds to John's creditor, Charles Alston. The order sent to Charles Alston but authorizing J. B. Hawkins reads as follows:

> Henderson Depot, Granville Co., N.C.
> Oct. 31st 1858
> Col. James B. Hawkins
> Matagorda Texas
> You will please have my Negro woman Nancy & child or children (wife of Cornelious) valued by some disinterested persons and deliverd to the order of Mr. C.J.P [Charles] Alston of N.C., which amount is to be applied to the credit of my part of the note due him, and oblige,
> Your brother,
> John D. Hawkins, Jr.[17]

Of all the letters, this order for appraisal and sale of Nancy and "child or children" is the most painful to come upon, effecting as it does the separation of a mother and her offspring from her husband and home on the Caney plantation. No situation more clearly shows the human cost of regarding persons as chattels to be sold.

But was Nancy really sold? Discovery of a list in J. B. Hawkins's pocket memorandum book dated September 13, 1864, six years after the sell order, suggests that she was not. J. B. Hawkins listed in pencil the names of fifty-two "people given shoes." Among the fifty-two slaves listed were Nancy and also her husband, Cornelious. This clue suggests that J. B. Hawkins himself paid his brother for Nancy and her child or children and that they continued living on the Caney plantation.[18]

The second way of thinking about slaves was an entirely personal one, speaking of them by name: Becky, Lucy, Osborne, Jimmy, or Aleck. Slaves had no other names than their first names until after the Civil War. Hawkins family letters show that to identify a particular slave, it was necessary to add a note on that person's history—for example, "William's boy Jim" or "bought of so and so." Genuine affection could develop on a plantation where work and home life intersected and daily interaction was personal. The work of women was especially open to this intimacy. Gardening, cooking, games, laughter, and childcare gave affection a chance. And the perils of nature—diseases, storms, or crop failures—were dangers that slaves and slave owners on a large farm faced together, however unequal their footing.

Ariella's letters to Mrs. Alston often included a personal message to or from a servant: "tell ole Mistress howdy," or "Becky sends howdy and a thousand thanks for her presents," or "Becky and Emily say give their love to all the white folks and Negroes too." And Ariella says to her mother, "Tell the Black people howdy for me." Personal relationships developed that were at odds with crass commercial calculation, yet the two ways of thought existed side by side.

The tenderest kind of personal caring is illustrated between Ariella and her long-time servant Becky during a cholera outbreak on Caney. Ariella's letter to her sister Missouri (Zuri) about the outbreak and Becky's illness is barely legible and bears no date. Ariella attributes the cholera outbreak to the fact that land had been cleared to the south of their plantation house and that decaying timber lay across this stretch. She held the common belief of the time that polluted air was the cause of cholera and chills and fever. She told her sister it might take a year or two for the environment to be less sickly.

> The Negroes are all healthy at this time. We had the cholera on the plantation before we left [for Matagorda]. It went entirely through the black family. Willis was the only one of the white family that had it. He was the first that had it. It is still quite bad in Texas . . . I hope it will never return for it is the most alarming disease (before you get used to it) you ever saw. I firmly believe if we had not of known how to have treated it and had the medicines ready prepared, we would have lost a good many of the Negroes. They were taken so suddenly and violently that I don't think they would live more than two or three hours. I thought we should lose poor Becky anyway. I was up with her all night. She would get easy for a little while and then have a [illegible]. She was perfectly well, had come in to get some medicine for one of the Negroes that had the cholera but before she had gone twenty steps from the door she was taken perfectly prostrate. Had to be carried to her bed. We gave her a hundred and twenty grains of calomel, a 100.20 grains of red pepper, 80 grains of camphor before she was relieved. She is perfectly well now and sends howdy to you all and says tell Dilce she received her letter and as soon as Sally is well enough she will send her an answer to it. The Negroes were all well last week. Mr. Hawkins's health is improving. Tell Mother I think her prescription cured him.[19]

A third way of referring to slaves as a group is by means of racial identity. By far the most frequent mentions in the family letters speak of slaves simply as "our Negroes." This phrasing sealed together the ideas of race and servitude. The word *slave* itself is hardly used in the Hawkins letters reported here. Its single use occurs in a letter by Ann Read Hawkins, writing of slave holders' "Christian duty" to their slaves. This duty, she implies, is their motive in leaving their North Carolina home and kin, for they must find land enough for their slaves to work in order to sustain their care.[20]

As the case of Edgar bears out, in the common assumption of the day, without enough land, or without the intention to work at farming, an already existing workforce would have to be sold, a ruinous prospect. Keeping the workforce and finding enough land for them to farm were among the factors that impelled James B. and Ariella Hawkins to move to Texas and Frank and Ann Hawkins to move to Mississippi.

Chapter 6

Building the Ranch House (Lake House), 1854

James B. Hawkins presented an impressive piece of news to the North Carolina family on January 12, 1854. To his mother-in-law he wrote: "I am very busy sawing out lumber for her [Ariella's] Lake Auston House. She is going to put up a large and splendid building and I hope after it is finished to have you to live with us. I think we will make a pretty place of it." By March 22, 1854, J.B. reported to Major Archibald Alston that the framing of the house was up.

> We will complete the frame of my Lake House this week. It is three stories high with nine rooms and cross passages and galleries all around with a large closet to every room and every room has a fire place. It will be a star house when completed. The sawmill makes lumber very fast. We are up to our shoulders in work with our different works to keep them all going as they ought to go.[1]

Ariella's diplomatic husband refers to the house as his wife's when he writes to her mother, but when he writes Major Arch, it becomes "my Lake House." J. B. Hawkins selected the location for the house and to a large extent oversaw the construction himself. He had the help of a master carpenter and the skilled craftsmen among the plantation's slave work-force. He supervised sawing the lumber cut from trees on his own Caney bottom land. The flooring wood in the house was ash. Some of the larger structural supporting timbers still had evidence of the bark. Even today, the marks of an adze are evident on some heavier beams, and nails in use had square heads. The house was certainly made from a detailed plan, but who drew the plan is still a mystery, although there are grounds for speculation.[2]

Possibly the design came from some printed, ready-made plan that skilled carpenters carried with them as part of the tools of their trade. At the time J. B. Hawkins had such a skilled carpenter working for him at Caney. Two years before the start of the Lake House, he wrote from Caney on January 19, 1852, that he had a ship's carpenter making a "vessel" for him (probably the schooner grounded in the storm of 1854, as described later in this chapter). J.B. also told Major Arch that he was having a "lot of fun" in a "pretty skiff" that the ship's carpenter had made for him. At the time of this letter, he was using the skiff to hunt ducks up and down Caney Creek and said they were having them every day for dinner. We do know from other house building in the county that ship's carpenters were employed in home construction.[3]

It was usual for plantations to have resident carpenters, blacksmiths, and brick masons. According to the census of 1850, James B. Hawkins had a carpenter named Nicholas Barr working for him at Caney. Barr was born in France, but he was probably doing routine carpentry on the plantation and was not the special "ship's carpenter" J.B. mentioned—that carpenter's work was a piece of news. Nicholas Barr, then, was probably not responsible for the Ranch House, although he may well have worked on it. The Matagorda County census of 1860 lists several carpenters but only two ship's carpenters. One was John Williams, born in Denmark and listed as a man of fifty in 1860; he would have been forty-four at the time the Ranch House was started. The other was Conrad Franz, who was twenty-nine years old at the time of the 1860 Census.

Details of the construction of the Hawkins Ranch House suggest that a ship's carpenter could have been instrumental in its design. The third story is held together in a "jib" construction, based on nautical construction methods including the creation of what is called a "pony wall," allowing openings through it. Evidence of this construction method is that the floors of the dormer windows are placed about two feet higher than the floors of the dormered bedrooms themselves.[4]

Why the house was built at all and why it was located in the middle of the prairie at the head of Lake Austin are questions never fully explained. Certainly J. B. Hawkins had an interest in creating an icon of his own success and in providing his wife and children with a much larger, finer, and more comfortable house than the plain one on Caney. He was not shy about saying he wanted to succeed as a man of business and even wanted to get rich. He teased Major Arch with a playful challenge: "I hope you will make a large crop, but I have my doubts about your making much money in N.C."[5]

Unquestionably, J. B. Hawkins took pleasure in building a house especially for his wife. She had told her mother that her husband "promises me many things" as soon as he began to produce sugar. Once the house on Lake Austin was built, J. B. and Ariella regarded it as their homestead, but it was not a permanent replacement for their Caney plantation. In part it was intended to be a better alternative to a rental house in Matagorda for their annual stays of several months. Having a home on Caney was still necessary for attending to plantation business.

Certainly the house on Lake Austin was a big improvement on the family home at Caney. The plantation house on Caney was the same house "bought of Quick" that had had to be patched up on their arrival in 1846. The new Lake House or Ranch House, as later generations called it, was two and a half times larger. In the Caney plantation house the living space was 2,488 square feet, with 920 square feet of porch space. In the new Ranch House, the living space was 6,428 square feet, with 2,232 square feet of porch space. The garden at the Ranch House was extensive, and when the yard was fenced off to enclose the house, oak trees, and orange trees, the enclosure was five acres.[6]

The end of the house on the Caney plantation came after years of serving as the home of J. B. Hawkins's youngest son Edgar and his family, to whom it was willed. In the 1940s the vacant plantation house was used briefly as a hay barn until it completely fell to pieces. In its day it was a two-story structure in a T-shape and had columns across the front gallery that reached up to the roof line. J.B. and Ariella referred to it as a dwelling that would serve their needs for the time being. The name "plantation house" encourages a false notion of grandeur.

But J. B. Hawkins intended the new house on Lake Austin to be grander than the one on Caney. It would be a "star of a house," he said. Besides pride in accomplishment and the pleasure of giving his wife a significant gift, there was another serious motive in choosing to build the Ranch House. Plantation families at Caney generally wanted to escape the chills and fevers that plagued them there. They put great store by the healthful effect of the sea breeze and of bathing in the Gulf of Mexico. J. B. Hawkins selected the location for the Ranch House in the open prairie at the head of Lake Austin in order to provide his family with the open sea breeze that came across Lake Austin from the Gulf. When there was any kind of outbreak of illness on the plantation, the family left Caney for the shore. They went to the town of Matagorda (then directly fronting Matagorda Bay) or to Sargent Beach, and when they were more ambitious, J.B. leased a vessel to take them to Pass Cavallo at the west end of Matagorda Peninsula, where according to his daughter Virginia, they got healthy and fat and bathed in the surf. The location of the new Ranch House would put the family closer to Matagorda than they were at Caney and would allow them to enjoy a commodious house open to the sea breeze.

It is true that the Lake Austin house was built in the middle of the prairie, far from the sight of any other house, and at first there was no grove of oak trees in front. Planting the oak trees was Ariella's highly successful landscaping effort. By today's standards the house seems isolated, built in a lonely place; there is no "neighborhood." But a big antebellum house was almost a neighborhood in itself. This house was for many people—children, visitors, relatives, servants, tradesmen, and those who came from miles away and simply showed up. Ariella's work was like that of an inn keeper or the head of a boarding school. One purpose of the house, both when it was built and ever after, was to offer hospitality. In its early days it housed guests for weeks at a time. Letters from the young Hawkins girls, Sallie and Virginia, written from "Lake Austin," show the liveliness of a sociable, fun-loving family. From various motives, then, the house began to rise on the coastal prairie in the year 1854.

As the family was beginning to enjoy growing prosperity, a disaster occurred that J. B. Hawkins could never have anticipated. In September 1854, six months after the framing of the Ranch House, the first locally recorded hurricane blew through Matagorda County, all but destroying the entire town of Matagorda and, farther east, striking at the Caney Creek plantation. An eyewitness, Mrs. Harris Bowie, described such a howling force of wind in Matagorda that her family had to flee, though nearly flattened in the wind as they ran, from one neighbor to another as houses were blown off their blocks and roofs were torn away. Homes, stores, and the Hotel Colorado House were all damaged or destroyed. The Episcopal Church (Christ Church, Matagorda) was blown down, and sailboats on Matagorda Bay were wrecked, including the schooner that J. B. Hawkins's ship's carpenter had built.[7]

When he reported the effects of the storm to Mrs. Alston, J. B. Hawkins said it did no damage to his plantation house on Caney but that it took down the big apse chimney on his sugar house and part of one gable end of it. There was such damage to his crops that he estimated making half the normal amount of sugar, about 250 or 300 hogsheads. He reported that none of the Negro houses were damaged on Caney, but much of his "best timber in the woods" was blown down. The town of Matagorda, he said, "was equally as bad as was represented." The kitchen of the new Ranch House was "blown flat." And a "Negro man, Sam, [was] crippled for life if he gets well." As a postscript for Mrs. Alston, he added that his schooner "is a reck on the Peninsula." But, characteristically thinking of what he must do next, he added, "I will get her off after a little."[8]

When the Ranch House was restored and completed after the hurricane damage, it became an increasingly lively place. Dancing lessons for the children every Friday and Saturday night and horseback rides across "our broad prairies" were family pleasures. Sallie and Virginia were close friends as well as sisters, and their letters written from the Ranch House give a glimpse of how it was in their day for young ladies to live there. Sallie wrote to her North Carolina cousin Jane mentioning the unexpected arrival at the Ranch House of their mutual friend, Mollie Duncan. Obviously Mollie had been to school in North Carolina with Sallie and Virginia and had made friends among their classmates and cousins there. Mollie's father, John Duncan,[9] had a plantation on upper Caney Creek in the northeast of Matagorda County. Sallie was twenty and Virginia (called Jennie or Jenny, and Sallie called her Little Sis) was seventeen.

August 24, 1857
Dear Jane

I received your most welcome letter last Monday the evening Mollie Duncan came. I had just finished reading it when I saw Capt Duncan's carriage drive up. Mollie looked very natural and you may know we were all delighted to see her.

Little Sis, Mollie and I set up laughing and talking until nearly three o'clock next morning. We got down to breakfast after everyone else had finished, and we had a house full of company too, but here everyone takes care of himself.

Mollie left yesterday for Matagorda to spend a week with her Aunt and cousin. Then will stay with us until October. We anticipate a gay Houston and I wish you would join us. There was a large [illegible] dinner and ball in Matagorda last week. I expect you think we do nothing here but frolic, which is very nearly the truth. Little Sis and I did not attend the ball as we had just returned from a party and felt too tired. You must write me all about your trip to school and all about the different girls. Dick Sommerville and Ell

Kennedy are both school mates of mine. I am surprised you should forget all about me so soon as to forget who my schoolmates were. I always thought Ell would be a belle, thought Dick would be very showy, but not stable, dignified. I expect her slightly as put-on. Little Sis says "she thinks Dick is coming out ask her if she has forgotten all her sadness at Patapsco." *We are living in a* progressive age, and a few years make quite a change. Who knows but what I may get pretty and dignified. Little Sis says "she is dignity personified."

Tell Aunt Zuri your writing is not her writing, that I hope you write on your own account, but I am glad she thinks of us sometimes. Give our very best love to her and all the family. I expect you will make an expert rider. You ought to have some of our broad prairies to gallop over. Little Sis and my pony had to run off this spring, and we have not seen them since. I am afraid we never will. Pa has been saying to get us some more but I had much rather have them [illegible] as they were beautiful horses and rode delightfully.

I'm surprised to hear that Mollie Battle did not like Warrenton and that she is so wild, very different from when I knew her. We have dancing school here every Friday and Saturday and have quite a house full on those days. We all enjoy it very much. Little Sis and I have a great deal of fun out of the little boys and girls. Tease them until we get them all crying then have to put our wits to work to get them quiet. Yesterday had the little boys crying because we told them the girls kissed them. Tell Bennet Johnson is the greatest from among the boys, he sends his best love to [illegible]. Ma Pa Little Sis join me in sending love to you, Aunt Bettie Cousin Bob Cousin Tom and all the children.

Write soon, don't wait as long as you did before. And ever believe me your sincerely attached cousin,

Sallie A. Hawkins[10]

Almost all the Hawkins children would be sent to North Carolina schools for their education before the Civil War. Just after the Civil War, Frank Hawkins would go to school in Germany and England. But illness and death claimed the lives of five of the nine children before they reached the age of twenty-five.

Chapter 7

Effects of Civil War and Emancipation

The two major events that marked the years 1861 to 1865, the Civil War and the Emancipation, affected the J. B. Hawkins plantation in opposite ways. During the Civil War the plantation thrived. But when the Emancipation took full effect (and it was not immediate), it ended the labor supply and the plantation system as well. Then the business of J. B. Hawkins changed from planting to raising cattle, and in that endeavor he would come to rely on his son, Frank Hawkins.

J. B. Hawkins's support of the Confederacy made him an even busier planter-merchant than he had ever been, because the demand for his sugar, molasses, cotton, beeves, lumber, and hides was now greater than ever. Texas did not suffer the same damage to its agricultural system during the Civil War as did the deep-south states that were overrun by Union foot soldiers. In Texas during the war years, slavery continued, as did the crop production in the plantation system.[1]

Sales receipts and other documents among the papers of J. B. Hawkins indicate that the Confederacy gave him favorable business opportunities. For example, J. W. Selkirk sent a notation on December 21, 1861, from Camp McCulloch, confirming a shipping order: "I enclose you an order on Col Hawkins for 30 BBLS molasses, 25 of which we have sold the Department [Confederate Department of Texas] at $24 per BBL. The other 5 we can sell here. I suppose the Col will not ask more than $16 as he promised us. Send it as soon as possible."[2]

J. B. Hawkins sent barrels of molasses and sugar to Port Lavaca by the paddle wheelers *Anna Catherine* and *Lizzy Lee,* instructing their captains to put his barrels as low in the vessel as possible. His notes in his memorandum book show sales of goods in addition to sugar and molasses. "For the government" there were shipments of leather, lumber, and hides that were sent "in vats" (of salt water). He made a note of 40 hides of #1 quality and 38 of #2 quality as well as four calf skins and eight goat skins. He had capital enough to buy cotton at the lower local price and transport it for sale at a higher price. His notes show that he bought 50 bales of cotton from Mrs. [Harris] Bowie and on September 15, 1863, sent "the government" 15 wagon loads and 12 carts. The carts he mentions were probably drawn by oxen. Many ox yokes are still stored in a barn at the Hawkins Ranch. Organized shipments like these were sent to Brownsville, then to waiting foreign ships off Mexican ports, and ultimately to the textile industries in England or other destinations. In this way the periodic blockades of the Gulf Coast were circumvented, and trade benefiting the Confederacy was sustained.[3]

Texas during the Civil War was made relatively safe from the fighting by its distance from the main battle fronts, but along its almost four-hundred-mile coast there was nervousness about a possible Union landing and march on Galveston. On October 4, 1862, the port of Galveston was taken by Union forces, but it was retaken on January 1, 1863, in a colorful combined sea and land battle. The Confederates planned their move on Galveston for New Year's Eve, when the attention of celebrants would be diverted by the holiday. Having few arms, heavy guns, or large ships, the Confederates made use of two river boats, the *Neptune* and the *Bayou City,* protecting their decks and gunwales with bales of cotton so that they became "cotton clads." Behind these bales they positioned crack sharpshooters who fired on Union sailors on the decks of more traditionally equipped craft, forcing them to seek cover below and opening the way to a boarding operation and the recapture of the port of Galveston. At the same time ground troops advanced from the mainland over a railroad bridge to secure the port for the Confederacy.[4]

Gen. John Bankhead Magruder, commander of the Confederate Department of Texas, had masterminded the retaking of Galveston, but he continued to worry about the coastal defenses of Texas and was intent on protecting Galveston. Following the contour of the Matagorda County coastline is a thin string of barrier islands and peninsulas often collectively called "the peninsula." This strip offered possible staging locations for Union forces if they intended to land and then march from west to east toward Galveston. To carry out such a plan Union troops would need to cross several creeks and rivers—Cedar Lake Creek (the border between Matagorda and Brazoria counties), the San Bernard River, and the Brazos River. For that reason in 1863–64 Magruder was especially active in locating troop encampments and fortifications to impede the crossing of these rivers in a move on Galveston. The general area between Cedar Lake Creek and the Brazos River was the "Gulf Prairie" region that Magruder filled with hundreds of troops located in about six campsites. The main encampments were located on the John Greenville McNeel plantation in Brazoria County. Some overflow troops from there were probably placed on J. B. Hawkins's land.[5]

Fort Caney, located at the mouth of Caney Creek, which J. B. Hawkins used to transport his sugar, was one of the places Magruder was intent on reinforcing; it was not more than ten miles south of the J. B. Hawkins plantation. Fort Caney was not one fort but the name of a group of four installations on both sides of the mouth of Caney Creek. One of the installations, on the east bank of Caney, was named Fort Hawkins for J. B. Hawkins. Fort Caney, Velasco (at the mouth of the Brazos), and Lavaca (Port Lavaca) were among places shelled by Union naval vessels. Lavaca and Indianola were occupied.

Fort Caney escaped occupation in the winter of 1863, perhaps because the gunboat was ordered elsewhere or because a Spanish sailing vessel, manned by Cubans, ran aground nearby and interrupted further Union attack. According to eyewitness Ralph Smith, the crew of the Spanish ship abandoned it and, as Smith said, "took to the woods." Then the men of Fort Caney helped themselves to the ship's cargo of coffee, potatoes, salt fish, calico, wash bowls and pitchers, and a mysterious tonic of "soothing syrup" they found "overpoweringly intoxicating," so that by midnight almost the "whole command" was stretched out on the sands of the beach, sleeping off the effects of the "soothing syrup."[6]

From 1863 to 1864, as General Magruder inspected his coastal defenses in "the Old Caney country," he stopped by to visit J. B. Hawkins at his plantation. The former governor of Texas, Francis Richard Lubbock, now serving as Magruder's adjutant, described the circumstance of the visit:

> General Magruder was very active in inspecting our lines and reconnoitering the movements of the enemy. With his staff, and sometimes a small escort, he was almost every day in the saddle, visiting our outposts to ascertain the strength of our positions and the spirit of the troops. In this way, early in December, we traversed the Old Caney country, stopping awhile at Hawkins'

> plantation and other hospitable places, and inspecting the works on the San Bernard. In returning we visited Velasco, everywhere finding along the front our gallant boys ready and eager for combat.[7]

To help with the fortification at the mouth of the Brazos River (Velasco), J. B. Hawkins sent one of his slaves named Aleck, who worked for sixty-seven days from August 10 to October 15, 1863, at one dollar per day. A copy of the receipt from the Confederate government for Aleck's work is on file in the records of the Hawkins Ranch. There is no family record of others being sent for such work, but very likely J. B. Hawkins sent more of his laborers without the expectation of being paid for their hire. Magruder's visit to the Hawkins plantation would have been an opportunity to request Hawkins's help.[8]

The Hawkins plantation at Caney, then, was not attacked by land or by coastal invasion, and the needed goods it produced for the Confederacy continued to flow, because slavery continued. The Civil War apparently did no harm to J. B. Hawkins's plantation and probably made it prosper.

The Emancipation, however, was a different matter. It effectively ended the plantation system that had remained productive during the Civil War years. Its end greatly benefited the slaves, although the benefits were a long time in coming and even then provided few opportunities for hired work other than in agriculture, where wages were low and housing was poor.

In Texas the Emancipation came on a special day, June 19, 1865, or "Juneteenth," a date later than either of President Lincoln's proclamations. Lincoln had made two proclamations of the end of slavery, both while the Civil War was being fought. His first, September 22, 1862, declared that all the slaves were now free in those seceding states that did not return to the Union by 1863. The second proclamation, on January 1, 1863, named the states to which the first proclamation applied. To proclaim is not, of course, the same as to put into effect; the actual freeing of slaves, wherever they were, came over a phased period as the Union forces advanced and added force to words.

In Texas the date that eventually freed the slaves came after Federal troops landed at Galveston, and Gen. Gordon Granger proclaimed on June 19, 1865, that "all slaves are free" and that the relationship between former slaves and their masters was now one "between employer and hired labor." The immediate effect of Juneteenth was that some former slaves—perhaps one-quarter, according to historian Randolph Campbell—left their plantations but that most continued to work in farming where they were already.[9]

Employment was limited for a people whose whole experience had been in agriculture, who had received no benefit of schooling, and who lived in a region of almost no industrial development. Many worked their own small subsistence plots and raised hogs, chickens, and crops. Some became tenant farmers for the old plantation owners who still had land. In the Emancipation, according to T. R. Fehrenbach, "more than 200,000 Negroes were cast adrift in one of the greatest social revolutions of all time. The first instinct of the plantation slave was to pick up and go. But he had nowhere to go." The initial joy of freedom was dampened by the realization that, as one freed slave put it, "freedom could make folks proud but it didn't make them rich."[10]

In order to make up for the loss of labor, one source indicates that for a time J. B. Hawkins contracted with the state prison system to lease convicts as laborers. The Census of June 14, 1880, reporting on Hawkins plantation residents, lists twenty-six "prisoners," all identified as "black male laborers." The convict lease system was sanctioned by the prison system of Texas until 1912, and the practice was not uncommon, especially for work in sugar production.[11] The fact that Frank Hawkins, J. B. Hawkins's son, is listed in the Census of 1880 as a "stock raiser" suggests that a shift toward cattle raising was occurring. Frank's younger brother Edgar ("Edker" on the census report) is listed as a grocer.

The Emancipation created a community of African Americans and their descendents, some of whom would become Hawkins Ranch tenant farmers and cowhands. Cowhands, especially, would be needed once Frank Hawkins arrived at adulthood and joined his father in developing the cattle business. In the view of Alwyn Barr, it has not been well appreciated that African Americans in Texas made up about 20 to 25 percent of cowboys, bronc riders, and cooks working in the cattle business during the late nineteenth century. The contribution of black cowboys has been overlooked, he says, because they have been omitted from twentieth-century novels about the West.[12]

I can confirm from my experience that the typical cowhand of Matagorda County from the 1930s through the 1970s was not an Anglo "Marlboro Man" or Hispanic vaquero but the African American. Frank Hawkins and later his daughters, the "young lady ranchers" who continued his cattle operation, relied on ranch hands drawn from this pool of people whose predecessors had been emancipated slaves. Chapter 25 features some of the African American cowhands of the Hawkins Ranch.

Chapter 8

Frank Hawkins and the Development of Cattle Ranching

Frank Hawkins did not join his father immediately in the cattle business. First he had to attend to his formal education. He was about nineteen in 1866 when the question of his education was raised in the family. His father discussed the matter with a family named Kirkland, whose son Jesse was about Frank's age. The Kirklands and the Hawkinses thought of sending the two boys together to a school in Germany. Both Jesse Kirkland's father and J. B. Hawkins likely got the idea of a German schooling for their boys from a German-American shipping agent both knew named Henry Runge of H. Runge and Company.[1]

The German school that Jesse and Frank attended was Carlshaven, located near Cassel (now spelled Kassel), Germany. That city lies in central Germany on the Fulda River southwest of Göttingen, in a Protestant region of the country. An ancient city chartered in 1198, Cassel was a refuge for Huguenots in the 1700s. In the early nineteenth century the Brothers Grimm lived in Cassel and collected their fairy tales in that region. Very likely the Texas boys Frank and Jesse had their passage arranged by Henry Runge and landed in Hamburg or Bremerhaven before they proceeded south to their school at Cassel.

Judging from the date of Frank's passport, the year their schooling began in Germany was 1866. Frank Hawkins's passport (in the family's possession) is signed by William H. Seward, the secretary of state under Lincoln. The passport is not a small booklet like today's documents but a single piece of vellum inscribed by hand in lines of spidery ink now turned brown. It is dated July 5, 1866, and is assigned the number 27209. There is no picture identification but only a verbal description:

Age: 16 years
Stature: 5 feet 8 inches
Forehead: high
Eyes: grey
Nose: short
Chin: oval
Hair: brown
Complexion: fair
Face: oval

It is puzzling that Frank's age in 1866 is listed on his passport as sixteen. We know from his mother's letter giving notice of his birth in Matagorda that he was born on December 17, 1847. His age in 1866, then, would have been nineteen, not sixteen.

At the school both boys were placed under the care of a housemother named Miss Meta Runge, who was almost certainly a relative of the German-born shipping agent in Texas. Miss Runge was a motherly lady whom Frank Hawkins gratefully remembered for her care. He named his eldest daughter (my mother) Meta, for Meta Runge, and when Miss Runge received the message that she had a namesake, she was full of happiness for the compliment and sent a small card from Cassel:

> Cassel, Germany Stande Place. 23. Feb 1892
> Dear Friend!
> Please, Mr. Frank! I read you are remembering for your "baby" Meta, and feel all the time very much love to you. Give little Meta so much Kisses from Old aunt Meta Runge[2]

Meta Hawkins was born on March 11, 1890, and Meta Runge sent her card two years after her namesake's birth. Perhaps Frank Hawkins sent her a Christmas greeting in 1892 and mentioned that he had named his baby daughter for her. When his daughter Meta was born, twenty-five years had gone by since his days as a student in Germany; still, he had fond memories of his student days there.

Frank Hawkins attended the German school for about three years, judging from a letter his sister Virginia wrote on July 14, 1869:

> We hear very often from Frank he is getting along very nicely with the Germans, seems from his letters perfectly satisfied. Ma & myself with Eddie [her brother Edgar] expected to go over to see Frank last June, but as Pa could not

> go with us, Ma gave out the trip until next Spring. I was very much disappointed for I had thought of nothing else for a long time. Pa expected to visit N.C. sometime this month, but he has abandoned the idea now as the yellow fever has broken out in Houston, Galveston, New Orleans in fact all along the coast & I fear they will have it in Matagorda though that place is now under strict quarantine.[3]

On completing his studies in Germany, probably in the fall of 1869, Frank went to a school in Warwickshire, England, in the village of Allesley, near Coventry. The school was called Allesley Park College. In 1848 Thomas Wyles and his wife Anne Ford Wyles had purchased a manor house there within a thirteenth-century deer park, and they converted the large, symmetrical three-story seventeenth-century house into a boys' boarding school. Frank boarded in this manor house. Whether his Texas friend Jesse Kirkland also attended is not known. In various written accounts about Frank Hawkins (his obituary, for example), the English school he attended is called "Ashton," but on his dance program showing the name of the school it is plainly Allesley, and one of his textbooks has the school's name stenciled in it. The dance card from Frank Hawkins's school years in England confirms my mother's memory that her father was a good dancer. The dance program is labeled on its front, "Allesley December 1869." There are eighteen remarkably varied dances announced, and F. Hawkins has a partner for each one. He penciled in each partner's name.

1. Polka	My Dear One	10. Mazurka	Miss Foord
2. Quadrille	Miss Foord	11. Lancers	Miss Willis
3. Galop	Miss Wyles	12. Croquet & Galop	Miss Florence
4. Lancers	Miss Wyles	13. Quadrille	Miss Flinn
5. Waltz	Miss Alice	14. Waltz	Miss Nellie
6. Croquet Dance	Miss Florence	15. Caledonian	Miss Flinn
7. Schottische	Miss Nellie	16. Schottische	Miss Wyles
8. Quadrille	Miss Willis	17. Lancers	Miss Dun
9. Waltz & Galop	Miss Alice	18. Waltz & Galop	Miss Nellie

Frank Hawkins was popular with the girls, and his school and nearby girls' schools organized dances to practice the social graces. Frank received a letter from the "Miss Willis" who danced the quadrille and the Lancers with him. From her school, Badminton House at Clifton (near Manchester, England), she wrote him an undated letter. Probably the Miss Willis who wrote and the Miss Wyles who is maligned with comic exaggeration in the letter conspired in writing Frank a girlish, flirtatious letter. Miss Florie Wyles, a student at Badminton House, was very likely the daughter of the headmaster of Frank's school. In reference to Florie Wyles and in collaboration with her, her classmate Miss Willis wrote:

> Badminton House
> Clifton
> Dear Mr. Hawkins
> I hope you will not believe any of the *foolish* messages that silly Miss Florie Wyles chooses to put in her absurd letters home ["home" being Allesley Park College] but we all know what she is. I hope you had some holly put in your bed the very first night you got back to Allesley [from a dance at the girls's school?]. Florie sends her *very very* best love & kisses.
> I remain yours truly
> M. E. Willis
> PS Emphatically not true

Frank Hawkins was at Allesley through 1870 at least. That year the headmaster authorized a charge to Frank's father from a clothier there.[4]

In the 1870s, when Frank returned to Texas after what must have been four years abroad as a student, approximately three in Germany and one in England, he turned to the cattle business and helped change the family enterprise from sugar growing to cattle raising. He worked hard to develop this new direction and to add new pasture land. A partnership in cattle raising between J. B. Hawkins and his son Frank now supplanted the earlier Hawkins plantation partnership of 1846. As J. B. Hawkins grew less active, the future of the Hawkins Ranch lay almost entirely in the hands of Frank Hawkins and, much later, in the hands of his five children after they came of age.

On November 23, 1887, at the John Rugeley plantation on Caney, a few miles north of the Hawkins plantation, Frank Hawkins married Elmore Rugeley, daughter of Dr. Henry Lowndes Rugeley and Elizabeth Elmore Rugeley. Frank was a man of forty; Elmore, born on July 2, 1867, was twenty.

To take up residence at the Hawkins Ranch House, the couple ordered a bedroom suite from Waddel's in Houston. It included a Renaissance Revival style bed with a high, ornate headboard and matching marble-topped tables, a wash basin, and dresser with a mirror. They made the Hawkins

Ranch House at Lake Austin their home and shared it with J.B. and Ariella, who were in residence there when not at the plantation house.[5]

The Hawkins Ranch House was the birthplace of all five of Frank and Elmore's children—Henry (Harry), Meta, Janie, Elizabeth (Lizzie), and Elmore (Sister). Harry, the eldest, was born in 1888, followed by a baby sister every other year until the youngest, Elmore, named for her mother and always called "Sister," arrived on April 3, 1896. Sister's birth and the immediate death of her mother is the beginning of the story of the young lady ranchers, who as children would move to town to live with their maternal grandparents, the Rugeleys, and who would eventually manage the Hawkins Ranch.

From the Hawkins Ranch House, Frank rode out with his cowhands periodically to gather up marketable calves and old cows for sale. Routinely he replenished his herd with purchased bulls and cows, a transaction that required writing to other cattlemen who had good breeding animals and making arrangements to pay for them. He had no means of mowing or planting grass to improve grazing, but he probably followed the custom of burning off the old grass at the end of winter to stimulate new growth. Setting fire to old coarse grass before spring had been a practice since the earliest days of Texas. So plentiful were wild game and small animals on the prairies that the fires sent deer, rabbits, and small animals running out of danger while vultures circled overhead.[6]

Frank Hawkins had no means of vaccinating or medically caring for his animals. He would have shrugged and counted it as fate to see a carcass here and there on a given day's ride. There were plank fences around the Ranch House and around branding pens but no fencing for interior pastures. Even external fencing was not generally used by ranchers in the county before barbed wire came into common use about 1884, a decade after the invention of the machinery to make it in 1874. Frank Hawkins, with the help of Liege Dennis, strung one of the first barbed wire fences in the county across the Sheppard Mott pasture. Barbed wire was so new that ranchers were skeptical as to whether cattle, able to see through the strands of wire, would be turned back by such a fence. To make sure, Frank Hawkins had the posts of his new barbed wire fence painted red. Fortunately, most of the external boundaries of the Hawkins Ranch, with the exception of the Currie segment and the rectangular Picket League that thrusts east beyond Liveoak Creek, were held in by the natural barriers of creeks and the shoreline of Lake Austin. Of course, in dry weather cattle might sometimes wade across a creek and mix with another rancher's herd, a possibility that made the cattle brand and earmark important.[7]

The seasonal work was to brand and mark calves, a task that required herding the cows and their calves through a narrow wooden chute, at the end of which a triangular gate allowed the smaller calves to pass into a branding pen, where a wood fire was kept burning and branding irons glowed red on the coals. From the 1860s onward the Hawkins Ranch brand was the H-Crook. Two or three ranch hands had to catch a calf, throw it to the ground, and kneel on top to keep it down while another pressed the hot branding iron onto its right hip. How long to hold the iron to the calf was a matter of art: just long enough to scar the hide in the pattern of the brand and not so long as to burn through to the flesh. Misjudgments were made.

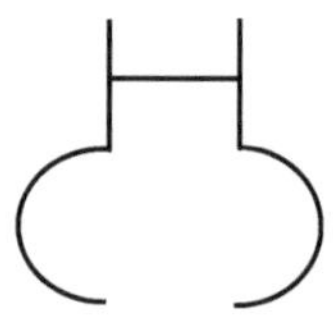

The brand of the Hawkins Ranch, the H-Crook.

Earmarks made bloody wounds. The Hawkins Ranch earmarks were "a swallow fork in the right ear and an upper bit in the left ear." In Frank Hawkins's day and the decades thereafter, the ranch hand in the branding pen held the calf down still longer and got out a sharp pocket knife to cut a large V out of the right ear and a smaller half circle out of the left ear. For many years afterward, these methods were the same. Castrating bull calves was done with the same sharp pocket knife, leaving a bleeding wound to heal on its own.

When Frank Hawkins's father came to Texas, beef was used for the family table but cattle ran freely, and as J. B. Hawkins once remarked, they were so wild that he would sooner take ten deer than one beef. In his day there was no fencing and he had few options for marketing or transporting calves for sale beyond the local area. In the earliest days, hides were more valuable than beef; stray cattle were routinely killed just for their hides, and their carcasses were left to the buzzards.[8]

Eventually three ways of marketing cattle became available to cattle raisers. The way on which so much cowboy romance turns involved taking the cattle on foot to market and grazing them along the way. Professional drovers gathered cattle from various ranches and branded them a second time with the drover's own herd identification for the passage. Then for almost a year the cattle were pushed along, allowed to graze and drink water in unfenced places known to experienced drovers. The stories of Matagorda County–born cowboy Charles Siringo are full of tales of the cattle drives to Kansas and other markets.

Another way of marketing cattle in the early days of Matagorda County was to ship them by boat. Charles Morgan was an entrepreneur who developed both rail and steamship lines to serve the ports of Indianola and Matagorda. His ships carried beeves and calves to New Orleans and elsewhere in the 1870s. Charles Siringo had a stint of working for the Morgan Line after being a stowaway on one of its ships. Frank Hawkins built cattle pens and a dock at Lake Austin. He may have shipped cattle by boat or shallow draft barge from that location.

After the coming of railroads at the turn of the century, cattle could be herded a short distance, held briefly in holding pens at the rail yard, and then loaded onto boxcars in Wadsworth and other points close by. Near Houston the Cassidy Commission Company and the Port City Stock Yards served to receive shipped cattle and managed the business transactions of their sale.

From sometime in the 1870s until his death in 1901, Frank Hawkins worked in the cattle business for his parents and his own children. As his parents aged, and after his younger brother Edgar's untimely death at thirty-six, Frank became the relied-on son. All who reported on his life and work praised his judgment and industry as a ranch manager, none more praising than his father. After Edgar's death in 1887, Frank Hawkins had no surviving siblings in Texas. His only surviving sibling, Virginia Hawkins Brodie, lived in Henderson, North Carolina.

As J. B. Hawkins became less active in the ranch partnership, Frank began to look to his wife's brother, Henry Rugeley, as a capable younger colleague in a brand new business venture. Instead of ranch land in the country, the town was becoming the center of economic opportunity. When Bay City was founded in 1894, supplanting Matagorda as the county seat, the new town invited new business projects. In 1898 Frank Hawkins and Henry Rugeley established a Bay City bank, of which Frank became president. It must have been a needed institution, because Texas was slow to develop banking service out of an antebellum "long standing bias against banking . . . and corporations."[9]

The fact that J. B. Hawkins's brothers often mentioned debts they owed to one another gives evidence of the lack of banking services in early Texas. Frank Hawkins's bank was organized as a private bank, which over the years went through several name changes: Bay City Bank, Bay City National Bank, and Bay City Bank and Trust. In 1903 it occupied a charming red two-story building of detailed brick masonry designed in a Victorian style, with arched windows and an arched recessed corner entry with one large column. The building still stands on the northwest corner of the courthouse square and received a historical marker in 1965.

Frank Hawkins's cattle operation and banking activity were highly credited achievements, and he had the further satisfaction of knowing he had won his father's admiration, as is expressed in his father's will. But sadness was to enter his life.

At the turn of the century, toward the end of his life, Frank Hawkins had to withstand three heavy blows. The first, the sudden death of his young wife Elmore on April 3, 1896, left him not only with deep sadness but the care of his young children. No forewarning prepared him for this terrible loss, if anything could have prepared him for it. A second blow fell almost immediately. On May 11, 1896, only one month after his wife's death, Frank's father J. B. Hawkins died at the Hawkins Ranch House at the age of eighty-three. The cause of his death, throat cancer, calls up a vision of the last, grim days when Ariella or a servant would come from the kitchen and climb the spiraling stairs to bring soft food and gently lift his head from the pillow. Although this death came with some forewarning, the foreknowledge hardly softened the double blow of losing both his wife and his father within a few weeks.

There was more anguish to come, and it would come from Frank's widowed mother Ariella, who began to make almost impossible demands of him. Yet they were demands that seemed to a conscientious man like Frank Hawkins to have merit.

Chapter 9

ARIELLA'S FIGHT FOR HER RIGHTS

One day about three years after her husband's death, Ariella took a good look at her husband's will. She understood immediately that it abridged her own property rights, and anger surged through this strong-minded woman. She found that her husband had simply dismissed from consideration her own community interest in land her husband had conveyed to their son Frank. In her life Ariella had always met the challenges that came her way: hard travel, settling in a new territory, loneliness, numerous childbirths, outbreaks of cholera and fevers, or the death of children. She was now approaching eighty, and nothing in her long experience had ever made her lie down in the face of difficulty, nor did she now. She became a passionate advocate for herself with a rhetorical power she directed against anyone she defined as an opponent—often against her own lawyers who were trying to look out for her interests.

J. B. Hawkins left to Ariella for her lifetime the Ranch House and the two hundred acres immediately surrounding it. But Frank received nearly all the Hawkins Ranch pasture land, excluding the Ranch House and the two hundred acres, which he would also receive at Ariella's death. Frank had been deeded land that his efficient cattle operations had made it possible to purchase. J. B. Hawkins was so beholden to his son for the piece-by-piece acquisition of this ranch land in the 1870s and 1880s, when his profitable sugar business was at an end, that he completely dismissed Ariella's community interest in it. As J. B. Hawkins explained in his will, it was

> because of the fact that for the last twenty-five years my said son has managed, saved for and controlled the said stock of cattle and by his constant attention and prudent management has worked them up to what they now are, and with the proceeds of each year's sale of cattle, out of said stock many of

> the tracts of land constituting my said pasture were purchased, and the titles taken in my own name and in many instances, and at divers times money derived from the sale out of said stock were advanced for expenses on my plantation and other property owned by me.[1]

This tribute reveals two things. First, the plantation business at Caney had not been paying a good return since the Civil War, and to some degree J. B. Hawkins had been dependent on his son's cattle operation to make up for a downturn in the planting business. Second, his gratitude associated with this ranching property led him to discount his wife's rightful community interest in it. Probably in the way of thinking that then and even later prevailed, J. B. Hawkins knew he could count on his son to take care of Ariella's needs and saw no reason to protect her rights.

But we can be very sure that Ariella herself did not discount her legal rights in the community property. She went into action, and beginning in the fall of 1899, a lawsuit was brewing. She sent a letter to the Angleton law firm of two brothers named Duff. At an agreed on date she climbed into her buggy at Caney and had herself driven to Velasco to the hotel. There she and R. C. Duff opened the question of her community rights. Duff immediately thought Ariella's case had merit, but he would need to consult the documents at the Matagorda County Courthouse in Bay City. That plan suited Ariella, and she would wait to hear how things stood after he had done "a very full investigation" that would take about a week's stay in Bay City.[2]

The attorneys were brothers F. J. and R. C. Duff of Angleton, Brazoria County, Texas. To get to the courthouse in Matagorda County, R. C. Duff went to an Angleton livery stable and hired a horse and buggy. After his investigations, Duff wrote to Ariella that she certainly did have a rightful claim and that they were "prepared to go into immediate proceedings for the establishment of your property rights." He probably filed the suit while he was in Bay City without immediately telling his client. When Ariella learned of this, she strongly objected to the action, which she thought too precipitous.

The Duffs were thoroughly professional and knew that a good way to force the acknowledgment of a rightful claim was simply to file a suit; then a settlement might be worked out if their client preferred. They told Ariella that they needed to hurry. Could she come to Velasco prepared to stay for a period at the hotel? They would need to take her deposition over a number of days. Ariella felt crowded into an action she was not yet sure she wanted to take: "I must see you before you proceed any farther so I may know something more about what I am doing." Nevertheless, for the moment Ariella put aside her doubts, and R. C. Duff visited her plantation on Caney and noted by memorandum that she "has this day [November 8, 1899] advanced us $200 with which to pay expenses of her suit against Mr. Frank Hawkins." As it would turn out, the two hundred dollars were to loom larger in Ariella's mind than all the acres in the Hawkins Ranch.[3]

Frank Hawkins was devastated to find himself named in a suit brought by his mother. He held the Duffs wholly responsible for pushing her into an action that might have been approached in a better way. His demeanor—a mixture of anger and sadness—is fully described by his own lawyer, Mr. Austin, by Mr. Duff, and by Ariella herself.[4]

Ariella then began to accuse her attorneys of acting in their own self-interest instead of hers. Her fury turned against her lawyers, and the two hundred dollars she had given the Duffs to start their legal investigations in her behalf now seemed to her an absolute swindle. She wrote, "You know well or better than I that you did nothing for my benefit rather an injury for I signed an instrument of writing by your advice which will debar me from ever being able to recover anything for my grandchildren after me so you will understand why I request you to place the same $200 to my credit in the Bank of Angleton."[5]

The Duffs were to receive compensation for their legal services if an acceptable outcome was obtained for Ariella either by suit or by settlement. If successful, they were to receive "between five and six hundred acres," as it was projected. They indicated that they would gladly take this measure of land in cash if she preferred.[6]

Ariella's fury only increased. The Duffs foolishly tried to respond to Ariella's blasts with facts and logic, armaments ever ineffective in the face of passion. For thirty-five dollars they had produced a deed for the land Frank Hawkins was to restore to Ariella and had had copies written by the county clerk, George Austin. For twenty dollars they had rented a horse, driver, and team to go to the courthouse in Bay City to examine the documents in the case. For five dollars they had purchased a revenue stamp to attach to the deed from Frank Hawkins conveying the disputed land to Ariella.[7]

Ariella wrote the Duffs a hot letter that they quoted in their reply. She had said, "Do you intend to pay the money borrowed from me, $200[?] If you do not you may be sorry for it, for if you place any value in your character as honest gentlemen or your reputation as lawyers, I shall most certainly make every effort to get it and shall not spare you one iota. The next you hear from me will be through the press."[8]

The Duffs were flummoxed by these rhetorical blasts from a client they were trying to serve and were smarting at Ariella's accusation that they, lawyers who prided themselves on their professionalism and honor, had cheated her and cared only about their own benefit. Writing for his younger brother, who was ill, F. J. Duff replied for the firm: "We have been in the practice some little time and have never before been accused of such a thing. . . . On account of the great respect I have always entertained for your late husband, Col. J. B. Hawkins, I regret this misunderstanding very much and am equally sorry that you have seen fit to so word your correspondence as to render any further negotiations impossible."[9]

Ariella's quarrel with her son had mutated into a quarrel with her own lawyers. Her complaints against them initially were only two, the fee and their unseemly haste in filing a suit. Later, when her lawyers were on the point of accepting for her a compromise settlement instead of pursuing a court case, she was infuriated that they had "mutilated" her case and by their incompetence deprived her of recovering her full community rights.

Ariella wanted the compromise to consist of awarding her the *whole* of the Duke League, an arrangement unacceptable to Frank Hawkins, who was distraught and unhappy to think of giving up so many acres that he counted on for the support of his children and, if need be, his mother.

Another compromise came into view in discussion among Ariella's lawyer Mr. Duff, Frank, and his lawyer Mr. Austin. Perhaps Ariella would agree to accept a portion of the Duke League instead of the whole? After discussing the matter briefly at Doubek's store at Hawkinsville, the three men went to present this possibility to Ariella at the plantation house, where despite being at daggers' points, she graciously prepared to provide them supper.

After supper a gentler scene ensued at the plantation house. Ariella began to soften and, in a quiet conference, asked Mr. Duff if he thought she was wrong to carry on the lawsuit against her son. He replied that he did not think she was wrong but that he did "deeply lament the remarks and feeling" she had expressed toward her son. Ariella, much affected by this observation, told Duff that "it was advice such as a son might have given." And she agreed to the compromise. The lawyers then went to the dining table and began to prepare deeds representing her acceptance of only a part of the Duke League instead of the whole of it.

But before they had finished, Ariella looked over at her son Frank, and seeing how distressed he was at the whole controversy, came quietly to Mr. Duff to whisper to him that "it seemed to hurt Mr. Hawkins so bad to give up the property, to give it *all* to him." Ariella's attitude toward her son now changed so much that she insisted on deeding him her entire interest, whether community or separate, in *all* the properties conveyed to him when her community rights had been ignored. She wanted to tear up the deed the attorneys had just prepared for her at her dining table but was advised that the law required a new deed. Asking for writing paper and pen, she then produced in her own hand a deed to Frank Hawkins of all her properties in the Hawkins Ranch, whether community or not. Once corrected and put in proper legal form, her deed was recorded and made effective on November 19, 1899.[10]

Frank Hawkins, after leaving the plantation and the contentious talk of lawsuits and settlements, began to appreciate the fact that his mother's community rights had not been recognized in his father's will. He addressed that injustice by buying his mother's interest for an agreed-on sum of cash that she gladly accepted and that was specified in the final, filed deed of November 19, 1899. Her attitude toward her son so changed that she insisted on deeding to him all her interest, whether community or separate, in all the properties conveyed to him on June 18, 1881. She also included in her conveyance to him the Ranch House and the two hundred acres surrounding it left to her during her lifetime by her husband.

The Duffs went away disheartened but not without a final proud word. In their view, their client had simply given away a right they had worked hard to have returned to her, and she had even deprived them of their fee. They wrote:

> If you have in the last settlement with Mr. Hawkins received a settlement fully to your satisfaction, you should not object to paying a fair fee to those through whose instrumentality and exertions you have received it. If you have gotten nothing, or if you do not think it just to pay for the labor that has been done in your behalf, kindly consider that you owe us nothing and accept the best wishes of a lawyer who has labored zealously and honestly in your behalf, and who while he errs (as all men do) in judgment, or when

> impugned, solaces himself with the consciousness of honest intentions and work as well performed as his capacity admits of.
>
> Yours very truly, R.C. Duff FJ & RC Duff[11]

During Ariella's correspondence with her lawyers in 1899 and 1900, she was in her late seventies. She held her own in the psychological warfare against her own lawyers and kept up spirited and punishing volleys to which they hardly had a reply. She knew intimately well every acre of land her husband had acquired during their long life together and was very clear about her own rights. But often she was simply angry and irrational. She lost the thread of the argument, took byways, and exaggerated the importance of small matters; but she never lapsed into helpless dependency.

There was one last unexpected turn in Ariella's life. She converted to Catholicism.[12] She died on March 9, 1902, and is buried beside her husband in the Hawkins family cemetery at the Caney plantation, surrounded by the gravestones of their children.

James Boyd Hawkins and his siblings at Raleigh, North Carolina, August 1887. Seated (left to right): James Boyd Hawkins, eldest of the group; Jane A. Hawkins; Dr. William J. Hawkins, addressed in several letters from James, especially about the building of the Ranch House; and Dr. Alexander Hawkins. Standing (left to right): Frank Hawkins, whose letters describe his trip with James from North Carolina to Mississippi, where Frank settled as James went to Texas—James named his son Frank for this brother; John D. Hawkins Jr., James's partner in the Caney sugar plantation; and Philemon B. Hawkins, one of several generations of Philemons. Alexander (Sandy) lived for several years in Florida and sent James orange trees sometime in the 1880s; the orange grove still thrives in 2013. HRLTD

Ariella Alston Hawkins, date and place unknown. HRLTD

Plantation house, painting by Georgia Mason Huston, about 1940. The house was in disrepair but still standing. HRLTD

The sugar mill at Caney plantation. The original 1850 painting by Sallie Hawkins, daughter of James B. and Ariella Hawkins, was copied in the 1940s by Georgia Mason Huston. HRLTD

July. Monday 15. 1861

Sept 13 1861

Given shoes To

Starlin

Shelby

Washington

Julian

Henry

Tuesday 16.

Bob

Kintchen

Amos

Abram

Carry

Ben Harper

Big Ben

Wednesday 17.

Williams

George

Peter

Lit Ezikle

Penney

Crawly

July, Thursday 18. 1861

Lit Ben

Caroline

Idellar

Beckey

Jane

Edmon

Charlett

Friday 19.

George

Laura

Emaline

Jesse

John

Cornelious

Saturday 20.

Francis

Amary

Horace

Shade

Jeff

Charly

James B. Hawkins carried with him a pocket memorandum book; several of his notebooks are in the Hawkins Ranch File of Historical Data. Here he has noted "people given shoes." HRLTD

A receipt from D. E. E. Braman for the hire of "Slave Ben," 1863. HRLTD

Matagorda Jany 26th 1863

Received of Col James B. Hawkins the sum of One hundred and twenty dollars, in Confederate treasury notes, the same being in full for hire of Slave Ben, the property of James Harper's heirs, from 1st Jany 1862 to 1 Jany 1863, one year.

D. E. E. Braman
Agent & attorney
for Jas Harper's heirs

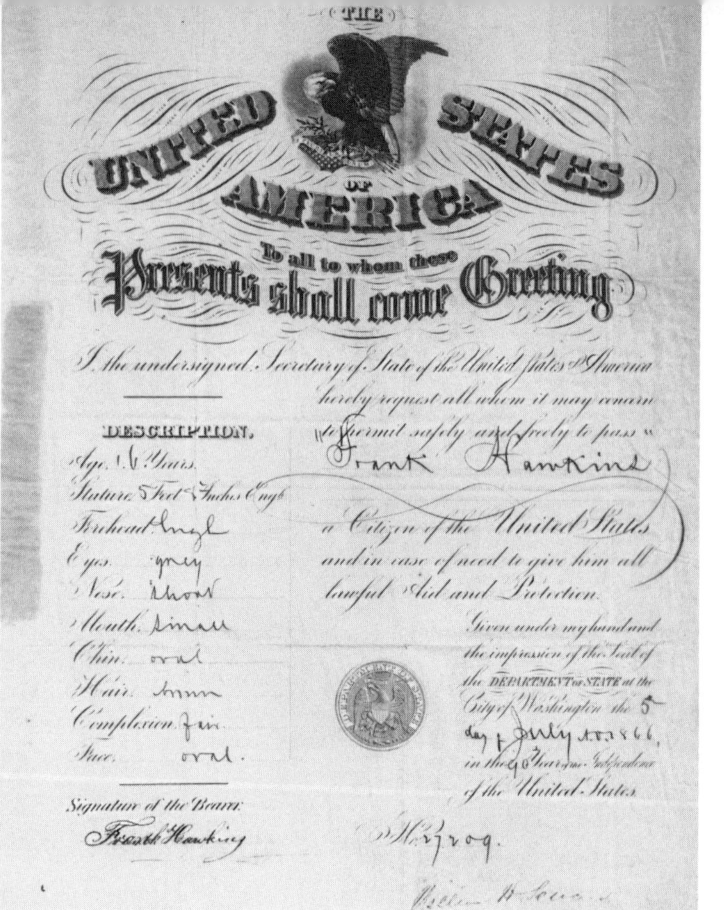
THE
UNITED STATES
OF
AMERICA

To all to whom these Presents shall come Greeting:

I, the undersigned, Secretary of State of the United States of America hereby request all whom it may concern to permit safely and freely to pass Frank Hawkins a Citizen of the United States and in case of need to give him all lawful Aid and Protection.

DESCRIPTION.

Age 16 Years.
Stature 5 Feet Inches Engl.
Forehead
Eyes grey
Nose short
Mouth small
Chin oval
Hair brown
Complexion fair
Face oval.

Signature of the Bearer.
Frank Hawkins

Given under my hand and the impression of the Seal of the DEPARTMENT OF STATE at the City of Washington the 5th day of July A.D. 1866, in the 90th Year of the Independence of the United States.

No. 27209.

William H. Seward

Passport of Frank Hawkins, issued on July 5, 1866, signed by William Seward. HRLTD

Frank Hawkins during his school years in Cassel, Germany. Photograph by G. Kegel, Cassel, Germany. HRLTD

Frank Hawkins, date unknown. HRLTD

Elmore Rugeley Hawkins, date unknown. HRLTD

Ruins of the plantation house, 2008. Photograph by Bill Isaacson, HRLTD

Part II

Young Lady Ranchers

At the beginning of the twentieth century, the story of the Hawkins Ranch hinged upon the generation of the five children of Frank and Elmore Rugeley Hawkins. After their marriage on November 23, 1887, Frank and Elmore lived at the Hawkins Ranch House, where their children were born. They were Henry Boyd (Harry), Meta, Janie, Elizabeth, and Elmore (Sister).

While still very young, the Hawkins children lost both their parents. Then they went to school and grew up in Bay City while living with their mother's parents, Dr. and Mrs. Henry Lowndes Rugeley. From the beginning the children were united by the shared experience of losing their parents. In 1917, in their twenties, they united further in asserting their right to manage the Hawkins Ranch themselves, displacing Henry Rugeley, their respected uncle and trustee. In the thirty years of their management, their sibling bond was strengthened through the livelihood they shared—and even more by their pleasure in being together in both town and country. Lizzie's disruption of the bond broke in like a betrayal. In 1921 she asked to have her interest in the Hawkins Ranch partitioned to her individually, and she withdrew from partnership with her siblings.

As Meta's daughter and the niece of the others, I knew them intimately for decades, and there is no part of my memory uninhabited by their presence. I describe them in personality and report the events of their lives that affected the history of the Hawkins Ranch: their mother's death and the cohesiveness derived from a shared loss; their schooling and the joy of having their own home; confronting their uncle to take charge of their ranch themselves; Lizzie's land partition, failure, and tragic withdrawal from them; and the decision to prevent the collapse of their birthplace, the Hawkins Ranch House.

Chapter 10

A Birth, a Death, and the Move to Town, 1896

Within a six-year period from 1896 to 1902, James B. Hawkins, Ariella, their son Frank, and his wife Elmore were all gone. Ariella's death in 1902 swept away the last but one of the Hawkinses who had come to Texas from North Carolina in the 1840s. Only Frank's sister, Virginia Hawkins Brodie of Henderson, North Carolina, was still living (see appendix for more about the antebellum children). The story of the Hawkins Ranch then continued through the lives of Frank and Elmore's five children.

The event that took these young children—my mother Meta and her four siblings—away from the Ranch House and brought them permanently to town occurred on April 3, 1896: the sudden death of their mother at the birth of their baby sister. The new baby was named Elmore for her mother but was always called "Sister."

Rowland Rugeley witnessed this event as a boy of seven. Even later in life it was a painful memory for him, but he described to me what happened.

"We were disturbed by the sudden arrival of a horseman who rode up to the house with an urgent message. It was Liege Dennis," Rowland said. "He told Mama and Papa 'Dr. Rugeley better come quick, Miss Elmore is very poorly.'"

Here was Dr. Rugeley being summoned to the bedside of his own daughter, who had just given birth but was now weakening dangerously. "We had to take the time to hitch up the horse to the hack," Rowland told me, capturing the family's sense of urgency and their frustration with the time it took to put the harness to horse and hack. Then they set off for the Hawkins Ranch House, going as fast as they could, "with Liege whipping up the horse" as they went. Rowland was taken along on the ride with his mother and father to provide the service of a small nimble boy. As they approached each of the several gates along the way, Rowland dropped off to the side, ran ahead to open it, and then caught up with the hack after Liege drove through and the gate was closed. Dr. Rugeley had anticipated that this help from his young son would speed their trip.

Dr. Rugeley arrived at the Ranch House and must have carried his black medical bag up the stairs to the west bedroom where his daughter lay dying. The hope was that if this revered physician attended his daughter, things would turn out well for her and the baby, but there was nothing whatever he could do. The baby was a healthy, beautiful girl, but her mother did not live.[1]

The mother's death occurred, according to the obituary, on the day following the delivery. No cause is stated in the obituary or in family papers. Many years after these events, my mother remembered how, as a small child, she had laughed when she heard her mother's cries. "Why I laughed I don't know. I just didn't understand, and I knew everyone was worried. . . . Children, I guess, can't understand such things and don't know how to act. The grown people had to tell me not to laugh. 'It hurts bad,' they said."

Dr. Rugeley was doubtless as well trained in medicine as anyone in the region at the time. He was a graduate of the University of North Carolina (1859) and of the Jefferson Medical College in Philadelphia, but what the medical profession had to offer in his day would have been scant. The unspeakably sad fact was that with only as much warning as a galloping horseman could bring, death took away his daughter, a young mother twenty-nine years old, who had been married only nine years earlier. Sadder still, her death left behind not only her newborn baby and her husband but four other very small children.[2]

Because there was such sadness associated with baby Elmore's name and birth, she was all her life called Sister. Although she was beloved, her birthday was not celebrated during her childhood because it was unseemly, the family elders thought, to celebrate on the day that had brought her mother's death. For her care, Sister was put into the arms of an African American wet nurse the family knew, whose baby had been born about the same time. Baby Elmore thrived and grew into a beautiful story-book picture of a little girl with long golden hair.

Far too young to look after themselves, too young even to know how to grieve, the Hawkins children were taken away from their Ranch House home and placed in the care of their Rugeley grandparents in town. Ever after, they would seek their comfort and companionship in one another and would feel that Rowland Rugeley, whose parents were their grandparents, was more a brother than an uncle.

In the few remaining years of his life their father, Frank Hawkins, made frequent trips to town on business or banking errands, sometimes stopping on his return to chat with a neighboring German family in the language he had learned at school in Germany. He saw his children at the Rugeleys' home when he went to town.

One of his errands in town may have been to obtain cash to pay his hands. A small one-room building (still standing) just to the west of the Ranch House yard was the place my mother remembered his using as an office for his payroll. He owned a big black safe, probably then located in this little building. The office had only one door, over which he had nailed up a good-luck horseshoe. Frank Hawkins is pictured standing in this doorway in a photograph from about 1899 (not included in this book). The structure, doorway and horseshoe are there to this day.[3]

Other family album photographs taken in 1899 (also not in the book) indicate that the children visited their father at the Ranch House on occasion. The pictures of one of these visits were taken by James H. Brodie, the son of Virginia Hawkins Brodie, when he was on holiday from the University of the South at Sewanee. Although they were separated geographically, a sense of fond kinship continued between Jimmy Brodie, his "Uncle Frank," and Jimmy's younger Texas cousins. In later years the Texas cousins made many visits to Jimmy and his wife Lucy in Henderson, North Carolina, and within my own memory he visited in Texas several times.

In the album the children look to be about the age they would have been when they had just started to school with Mrs. Holmes, their first teacher in Bay City. In one photo Frank Hawkins is shown with them and their first cousin Ella Hawkins, daughter of their deceased uncle, Edgar Hawkins. Frank is seated on the west steps of the Ranch House front porch, facing the afternoon sun. He wears no tie; his shirt is buttoned at the neck and he wears a suit jacket with a vest and a dark felt hat. His trousers are stuffed into his boots, which are tall, dusty, and well worn, and they have long leather boot pulls that droop from their tops like a spaniel's ears. He has an impressive mustache that curves downward on each side of his mouth, and he holds in his uplifted right hand something that looks like a cigar or cigarette. He looks exactly like a character out of Chekhov's Russia.

In the 1940s when these children of Frank Hawkins were fully in charge of the Hawkins Ranch, they remembered their grandmother Ariella but could not have known her well. Sometimes when I was with them at the Ranch House exploring some corner of the place, we would come upon a reminder of Ariella and J.B.—"Oh Margaret, here is J.B.'s own branding iron." Or "Oh, look. Here's Ariella's side-saddle. You want to try it out?" To me it was a curious looking piece of leather with just one stirrup and, like English saddles, no pommel. It was surprisingly comfortable to have only the left foot down in the stirrup and the right knee clamped around a leather-padded hook.

As children my mother and aunts were too young to know much about family life and recreations in the early days of the plantation on Caney or at the Hawkins Ranch House. They had no knowledge of Ariella's beautiful sorrel named Red Bird or how Sallie and Jenny used to love to ride across the broad prairie, turning their complexions too brown despite wearing "our bonnets," as they assured their grandmother. My mother and aunts would have been delighted to know of the dancing lessons Ariella arranged for her children at the Ranch House.

On the third story of the Hawkins Ranch House, in a sadly broken state, was a diorama that Ariella had made. On a board she had painted a pale blue background and glued onto it sea shells and grass she had gathered at the Gulf shore. She had her creation framed in a now broken glass box. Ariella's buggy was still kept in the barn. The black leather tufted upholstery was scarred and had some protruding springs, but in the late 1930s and early 1940s, my family would sometimes have it hitched up for a hunting or pecan gathering expedition, and nothing was more fun.

Having shared childhood insecurity, the Hawkins children ever after found their safety in one another. Their strong family alliance and genuine enjoyment of one another meant that they would find any family member's disaffection extraordinarily painful. Lizzie's eventual estrangement from them would sadden them terribly. The loss of April 1896 meant that the Hawkinses in years to come would live in town and manage their ranch from there. It also meant that Sister, the motherless newborn, would always be loved with special tenderness, would become a warm and loving person herself, and would always feel especially close to Rowland, who rode along with his parents on the day she was born and her mother, his sister, lay dying.

Chapter 11

Schooling and a House of Their Own, 1913

Living in town with the Rugeleys, the young Hawkins children had only to cross the street to attend classes with Mrs. J. D. Holmes, who was assisted from time to time by her daughter, Miss Tenie Holmes, who many years later would be my teacher. Mrs. Holmes was a widow who had written for newspapers when she lived in Kansas and was highly regarded by her friends for her "always instructive conversation."[1]

When Mrs. Holmes wished to share one of these instructive thoughts, it seemed natural for her to write a letter to the editor of the local newspaper. In 1899, and again in 1900, the Hawkins children had given her a Christmas gift. She was touched and filled with high-minded resolve to teach them well. She wrote:

> Mr. Editor:
>
> It is said that 'old wine is best to drink, old wood to burn, and old friends to love,' and I may add that old thanks are sweetest.
>
> December 24th, 1899
>
> *'Twas the night before Christmas,*
> *And all through the house*
> *Not a creature was stirring,*
> *Not even a mouse.'*
>
> When the spirit of Santa Claus embodied in five little spirits by earthly names of Ella, Meta, Janie, Harry and Rowland stole silently (at the witching time of night) into my dining room and placed on the table a beautiful silver syrup pitcher containing a note expressive of love and respect. I appreciate the sweetness of the motive far more than I ever could the material juice of the cane. It is the little things of life that count in making up the sum total of happiness.

December 24th, 1900, 4:00 a.m.

I arose early Christmas morning bent on early and suitable preparations for the day. Passing through the hall my attention was arrested by a handsome eight day clock which evidently arrived in the same mysterious manner as on the previous Christmas even and the same little people were the donors. When I wound it and started the pendulum, I found that it was not striking correctly. I rather feared to tamper with it lest I might injure the mechanism, and then this overwhelming thought came over me. I have sole care of the mental mechanism of five bright, willing and teachable children and if by precept or example of mine they should go wrong in life how great would be my condemnation. The clock has taught me a lesson.

Mrs. J. D. Holmes, Teacher[2]

When the Hawkins children had completed all the classes that Mrs. Holmes offered, the older ones—Harry, Meta, and Janie—went on to study with the Reverend John Sloan, rector of St. Mark's Episcopal Church. Sister and Lizzie, when they were old enough, went to the local public school. To his pupils Father Sloan was a revered figure. What he taught had authority for his pupils, and they remembered him as a man dedicated to scholarship who held his students to a standard of diligence and good manners by which they learned to measure themselves. He wore a clerical collar and was a slender, bespectacled man with merriment in his countenance. His social gifts made him good company.

Sometime just before 1907, when Rowland was about eighteen, he wanted to apply to the University of Texas and decided that taking Mr. Sloan along with him and appearing in person would strengthen his case. Off the two of them went to Austin, straight to the office of the president of the university. In later years Rowland enjoyed quoting what Mr. Sloan said to the president as they were received: "This young man is coming to you in nothing more than the raw state of nature."

When Rowland's record was examined, he was advised to prepare for entry at an Austin academy and then apply to the University of Texas. He followed the advice, was then admitted in 1907, and graduated in law with an LL.B. on June 11, 1912. Probably it was in law school that Rowland learned to value evidence as the means of verifying statements, and when good manners or a teasing spirit permitted it, he would test what he heard said by calling for the supporting facts. He did not like to dramatize; he preferred facts, but good humor was never absent.

When Meta, Janie, and Lizzie finished with local schools, they went as boarders to Kidd-Key College in Sherman, Texas. Meta graduated there in 1909, Lizzie in 1912. Janie attended but probably did not stay to graduate. Kidd-Key was a well-known boarding school at the turn of the century and drew young ladies from all over Texas in its day. The school offered the standard courses in history, English, chemistry, and mathematics. It also offered an ambitious music curriculum. Students worked up recitals for instrumental and vocal music, and faculty members offered duo piano performances of Tchaikovsky, Chopin, and Liszt.

Annual junior-senior formal dinners were given at the Binkley Hotel in Sherman, and Washington's Birthday was another annual celebration that rated a dinner at the Binkley. Students often planned informal get-togethers in their rooms. Appealing little hand-drawn invitations were pushed under a student's door: "If you have nothing to do, drop in room 59 Friday evening about 8:30 with a mug and a spoon. . . . Hot chocolate, sandwiches (potted ham preserves) baked beans."

Meta's Kidd-Key album had a red leather cover stamped in gold with the title "The Girl Graduate, Her Own Book, Meta Hawkins, A.B. 1909." In it her "prophecy" is given by another student: "You will go across the water (the Potomac) and you will meet a fair man who will be your affinity." By the fall of 1909 Meta had graduated from Kidd-Key and was a student at Chevy Chase College in Washington, DC. She was the only one of her sisters to attend there and may have stayed only one year.[3]

About 1912 Sister was ready to go off to school. Probably because Rowland and Lizzie were already in Austin, she chose that location instead of Kidd-Key. Rowland went with Sister to Austin to talk with Miss Mary Whitis about entering the boarding school named Whitis, established by Mary and her sister, Gertrude. It was at the corner of 27th Street and (now) Whitis Avenue, the present location of the Scottish Rite Dormitory. Both ladies were college graduates and sought to found a school to prepare students for a college or university curriculum. They thought the public high schools throughout Texas were not uniformly equipped to prepare students for this level of study.

Whitis School placed its announcement in the newspaper each year: "High standard college preparatory boarding school for girls. Certificate of graduation admits students to Texas State University and other universities without examination. Also to Wellesley, Vassar and other eastern colleges.

Kindergarten, primary and college preparation." Whitis offered courses in "English, Mathematics, History, Physiology, Geography, Civics, Latin, German, French, Spanish, Music, Art, and Physical Training." No class would be too large, the school promised. Latin was especially emphasized. Extra courses for an extra fee were available in piano and art.[4]

The school did admit boys, but only girls were boarders. The girls were expected to do necessary shopping only on "suitable occasions" and were to attend lectures, plays, and concerts with a chaperone. On the first Saturday evening of the month, the girls were "formally at home to their friends." Parents were urged to give their daughters a stated allowance "in order that the girls may learn the value of money." Physical training was compulsory, and each girl was expected to bring her own "gymnasium suit, made of black serge, in divided skirt pattern." The cost for tuition and board and laundry was $435 per year.

As Rowland and Sister sat across from Miss Whitis, conferring about Sister's entering the school, Rowland, then a law student, began to worry about Sister's obligation to pay the tuition if, after a few weeks, she chose not to stay. Knowing Sister, Rowland thought she might not like all that Latin, and the mathematics and history, and the physical training while wearing a black serge divided skirt. What would happen if sister got into Whitis but found that she wanted to come home? On Sister's behalf, Rowland strenuously probed the tuition issue with Miss Whitis. Years later, he remembered his face-off with her: "Miss Whitis and I just went round and round on the question of what would happen to Sister's tuition if she didn't like the school and came home. I didn't want her to have to pay all the tuition if she got homesick before the term was up."

Sister did stay and found that at Whitis there was also instruction in music, the arts, and the social graces. At the annual May Fete, costumed dancers wove colorful ribbons around the long maypole, and there were numerous piano recitals. At one, held at Gregg House on 27th Street on May 27, 1913, Elmore Hawkins was on the program to play "A Shepherd's Tale" by Nevin and "Waltz" by Karganoff. Sister was seventeen.[5]

One other piece of evidence from Sister's time at Whitis is a thick book of the selected essays of Thomas Babington Macaulay. It was required reading for second year English students. On a blank page at the front of the book is a pencil sketch of a young lady's profile, nicely drawn in Gibson Girl fashion, and an exchange of notes with another student:

You know that candy we have in our room. Everybody that comes in just helps themselves to it. So you come get some of it."

I'm just hipped on Henry . . .

You get hipped on the funniest people.

When Lizzie entered the University of Texas after graduating from Kidd-Key in 1912, she was a spirited beauty who had many friends and beaus. She affiliated with the sorority Kappa Alpha Theta and stands in that group's picture in the university's annual, the *Cactus,* in both 1913 and 1914. She also joined the Rabbit Foot Club, the purpose of which was to give young ladies the opportunity to organize dances like those of the young men's German Club. Lizzie's course of study included English, French, and history. Had she stayed, her graduation year would have been 1916. Two of her Theta friends, Elaine Lewis and her distant cousin de Rugeley Peareson, were standing with her in student organization photos for Theta and the Rabbit Foot Club. Very likely Lizzie and Rowland initiated social engagements for each other while they were at the university. Certainly Rowland knew and went out with Lizzie's friend Elaine Lewis; he described to me his mad dash, with several mechanical failures, on a borrowed motorcycle from Kyle to Austin to return in time for a date. "My hair was filled with sand and grit," he said.

About 1912 or 1913, the children's uncle and trustee Henry Rugeley began to give some thought to their coming of age. Harry was then twenty-five, Meta and Janie were twenty-three and twenty-one, and Lizzie and Sister were nineteen and seventeen. If they wanted a home of their own now, they were entitled to it, and they could afford it using the trust that their father had established for them. Their uncle, as trustee, could certainly justify spending money from their trust to build a house in Bay City for the beneficiaries, now almost grown and without a home of their own. My mother told me that their uncle did not happen to think to give them any choice in the selecting of material, style, or size of the house. He assumed his responsibility was to choose for them. He was not more than forty years old, but as was not uncommon, his sense of responsibility took a paternalistic turn that unwittingly narrowed the choices of those in his care. That fact, however, did not in the least dampen the joy of all five Hawkins children when they moved into their new house in Bay City, and their uncle's choices have stood the test of time. He paid out of their trust a total of $37,009.70 for the purchase of three lots, the construction of a three-story brick house, outbuildings, and furnishings for the house.[6]

This house, which in my childhood came to be called "Aunt Janie's" because she lived in it for her lifetime, was the first home of their very own after the Hawkins children left their Ranch House. As soon as they had installed their dark oak Renaissance-style furniture with spiral legs and had hung the pictures, they went out and got a professional photographer to take pictures of the interior and of themselves. To record their happy new beginning, they posed themselves at a time of day when light beamed onto the south screened porch.

One of the first things the Hawkinses faced as new occupants of their own home was the Colorado River flood of 1913, a time when "the waters of the Colorado and Brazos rivers met." Local people talked about the "raft" of logs and debris that blocked the flow of the river and that extended for miles upstream from Bay City. Attempts had been made to channel around the raft, but they were largely ineffective, and during the frequent floods the people had to contend with moving through the streets in rowboats. Sister was photographed in front of their new house, sitting alone in a rowboat, manning the oars on Avenue G, the Boulevard. The streets were lakes of water, but the sidewalk in front of their house and the house itself were clear. Young people seemed to take these frequent floods in stride. Word would go around that they should all gather at the bandstand on the courthouse square, and they would come to sit in their rowboats to hear a concert of the local band.[7]

Sometime in 1914 the Methodists scheduled a big conference in Bay City and asked the young Hawkinses to provide hospitality to an important delegate. Meta ever after laughed about how Lizzie had insisted that their visiting dignitary must have just one more cup of cocoa. "No thank you," said he. "No, I really just couldn't. Thank you, but no." But Lizzie would not allow their guest to curtail the hospitality she was intent on heaping on him. Overriding his protests, she sent to the kitchen for the cocoa. Then back came the deflating report that there was no more.

Their guest, a Methodist bishop from Birmingham, Alabama, wrote a gracious letter of thanks on November 25, 1914: "I beg to be permitted to express again my grateful appreciation of hospitality of your home during the session of the Texas Conference and to tell you that your many kindnesses brought me under lasting obligations. I have never been more delightfully entertained."[8]

On April 3, 1917, the youngest of these young ladies, Sister, would turn twenty-one. Her coming of age brought a difficult decision, one they talked over together for some time before deciding what they must do. Should they attempt to run their ranch themselves or allow their Uncle Henry to continue managing it for them?

Chapter 12

Young Lady Ranchers in Charge, 1917

When the Hawkins children were still very young, still living with their grandparents and attending Mrs. Holmes's classes, their father, Frank Hawkins, had begun to be troubled with symptoms indicating a kidney ailment called Bright's disease. His father-in-law Dr. Rugeley had suggested that he seek treatment in Austin with a Dr. Wooton. Dr. Rugeley accompanied him there, and the two of them stayed at the Millet Mansion in downtown Austin. On February 25, 1901, while in Austin, Frank Hawkins died suddenly, and his remains were brought by train to the small neighboring town of Van Vleck, where he was buried beside his wife at the old Rugeley cemetery. He was fifty-four years old, and his children were still very young. Harry, the eldest, was only thirteen and Sister, the youngest, was five. They would continue to live with their Rugeley grandparents in Bay City.

As Frank's will provided, a trust for his minor children was soon put into effect. Henry Rugeley and James H. Brodie were both named as trustees, but Henry Rugeley, as the local resident, was the more active of the two. He began operating the Hawkins Ranch for his nephew and nieces in 1902 and continued to do so for fifteen years. He was scrupulously businesslike, as the records of his bookkeeper Frank A. Bates attest. A statement of cash receipts and disbursements for May 1915 reveals the ranch operations of that period. The report indicates that a sale of cattle on May 4, 1915, brought $7,495 and that the number of calves branded in the spring was 654. A notation also shows that the Brahma breed was then being recognized for its value in improving native herds.[1]

Frank Hawkins's will provided that the trust should continue until the youngest of the children had reached her majority in 1917. At that time the trust benefiting all five children would terminate, and the properties would come to them free of trust.

For a reason we do not know, Frank Hawkins had willed to Harry Hawkins alone that part of the Hawkins Ranch called the Sheppard Mott, the northernmost pasture of about 3,300 acres. ("Mott" indicates a clump of plainly visible trees in the middle of a prairie. "Sheppard," often misspelled Shepherd, is the name of the early settler Abram Sheppard, from whom J. B. Hawkins bought land.) This Sheppard Mott land was to be held in trust until Harry reached twenty-one. All the rest of the Hawkins Ranch (25,431 acres at the time of the will) Frank Hawkins left to all five of his children, Harry included. Harry turned twenty-one in 1909, eight years before his youngest sister reached her twenty-first birthday. Henry Rugeley stated in a court record of 1917 that Harry's trust was ended and his property was transferred to him in 1909, near the date of his twenty-first birthday, August 21, 1909.

Harry was born with a disability, perhaps cerebral palsy, that made him lame and diminished his capacity to take complete charge of his own business affairs. Given this fact, Harry's taking delivery of his inheritance free of trust did not mean that he assumed the active management of his land and cattle. Henry Rugeley kept his books in good order to reflect Harry's ownership, and Harry signed papers presented to him from time to time. But Henry Rugeley would have felt it a duty to continue managing Harry's part of his father's land and cattle just as he was doing with the rest of the Hawkins Ranch still held in trust for all five of the children. There was little perceptible difference in Henry Rugeley's active management of property whether it was held in trust or free of trust.

The young Hawkins siblings, now living all together in their new house in Bay City, were thoroughly familiar with the terms of their father's will and talked among themselves of its implications. They knew very well that when Sister turned twenty-one they had a legal right to their property free of trust, but they did not know exactly how to assert that right. They knew, however, that if they did not say they intended to manage the ranch themselves, matters would continue as they had for fifteen years. They discussed with one another how they would manage the ranch and thought they could do it successfully. They worried that the day of Sister's majority would come and go without their uncle's recognizing any change. They thought that in a conscientious effort to take care of them, their uncle would continue the management of their property, as he had in Harry's case, even after the termination of that trust.

Their uncle would doubtless have assumed that without him, these young ladies were not capable of managing their property themselves. Henry Rugeley also pastured some of his own cattle with theirs on their land. Of course, he branded their calves with their H-Crook brand and his own with his own brand. But the young Hawkinses, talking matters over together, thought mixing cattle from two different herds was not the best practice. They thought they should fully stock their own land with their own cattle.

Year by year, as Sister's twenty-first birthday approached, the Hawkins siblings consulted one another about how to divide up the work of running the ranch "when we take it over." On the date of her birthday, Sister was to write a letter to Uncle Henry to tell him she had now turned twenty-one, and Meta was to go to the bank to break the news to him. Sister many times described to me the family deliberations as they prepared for an awkward confrontation with their beloved uncle. From my memory, her description ran like this:

> Meta said when we took over that she would be able to keep the books all right, and she could see to the farming of cotton and corn over on Liveoak Creek. And Janie said that she could see about the cattle. We were thankful to have Dode Green on the place as foreman, and we knew we could count on him and the cow hands he could get hold of when the cattle needed working. Dode was a tremendous help to us. As the youngest in the family, when I turned twenty-one on April 3, 1917, I wrote a letter to Uncle Henry reminding him that I, the youngest, was now twenty-one and no longer a minor. Papa had put it in his will that the guardianship was to last until the youngest reached twenty-one. When my birthday came, Meta said she would go to Uncle Henry and just put it up to him that we would take over the ranch. He was flabbergasted and, of course, concerned that we were being very foolish.

They were betting their whole future on their own untested capacity to run a ranch. They were four young women taking over from a seasoned businessman, and the business world, especially the ranching world, was a man's world. It took not only courage but the surmounting of a kind of apologetic embarrassment that invaded their spirits as they faced their uncle with such an unwelcome and seemingly ungrateful proposal. Very likely their uncle had assumed that for their sakes he would run their ranch for them always, and perhaps he thought that in the absence of a fee, pasturing some of his own cattle was reasonable.

The confrontation was without any feeling of hostility. Tearing at their self-confidence was their unwillingness to disappoint their revered uncle or to seem foolish or disrespectful. Nevertheless, Meta went to Uncle Henry to say simply that since all the Hawkins siblings had come of age, they would now take over the management of the ranch themselves.

Certainly their action shocked their uncle and shocked Rowland as well. He so admired his older brother Henry that it would have seemed to him the best possible arrangement to have Henry continue to run the Hawkins Ranch.

What happened next was clear-headed and businesslike. Henry Rugeley and his co-trustee James H. Brodie petitioned the District Court in Matagorda County to examine the trustees' management of the Frank Hawkins trust, to "pass upon the propriety and correctness of trustees management," to discharge the trustees of further liability as they transferred the properties to the beneficiaries of the trust, and to advise the trustees of a method by which to make the transfer. The trustees wondered, for example, whether any allowance should be made for having paid out slightly more income to some beneficiaries than to others.

Bookkeeping entries for 1915 indicate that Meta, Janie, and Elmore were each paid an allowance of two hundred dollars per month but that Lizzie was paid three hundred dollars per month. The difference may have been a recognition of Lizzie's household expense: she had just been married in 1914 and was the only sister then married. That no allowance expense is indicated for Harry probably reflects the fact that his account had been in his own name since 1909, and he signed his own checks.

The trustees also asked the court for guidance on whether the property was to be divided and distributed to each beneficiary separately or kept as a unit and distributed to them in partnership. As things turned out, the Hawkinses kept the property as a unit and operated as a partnership.

In their document the trustees state to the court how much land they, as trustees, had received to manage in 1902 at the time their trusteeship began. The land received by trustees was 25,431 acres, then valued at $75,801 ($2.98 per acre). Land that the trustees now transferred to the beneficiaries in 1917 was 26,922 acres, valued at $269,220 ($10 per acre). The increase of 1,491 acres in the land inventory represents a purchase made by the trustees of contiguous acreage used fully, the trustees said, as a part of the beneficiaries' pasture. Also mentioned was the purchase of three lots in Bay City and construction of the brick dwelling and outbuildings—all paid from the trust at a cost of $37,009.70. This provided "a home for the comfort and enjoyment of the beneficiaries," and the trustees felt the expenditure was justified.

In summary, the trustees asserted that in 1902 at the inception of the trust, they had received a total value of $140,728.11 (in land, cattle, horses, hogs, wagons, furniture, and cash). Now, in 1917, at the close of the trust after fifteen years, the total value of the trust property was put at $474,669.06.

The trustees also stated to the court that the Sheppard Mott, left in trust to Harry Hawkins, had been transferred to him outright at his majority in 1909. Thus the young lady ranchers, on their own now, would be managing a 26,922-acre ranch, to which Harry's more or less 3,300 acres of Sheppard Mott land would be added, making them responsible for a ranch of about 30,000 acres.[2]

Chapter 13

COURTSHIP AND MARRIAGE

Although living in their own house with their brother, the Hawkins girls in their social engagements were under the scrutiny of their Rugeley grandmother until her death in 1923. Sister, growing up with her grandmother, described her as an exceedingly gregarious person. "Just sit up with me a little while longer," she said to Sister, whose school girl eyes grew heavy at a late hour. "Let's just play one more hand of whist." Though sociable, Mrs. Rugeley was also very conscious of propriety, and her standards derived from a previous generation. The fact that the Hawkins children skipped their parents' generation to be brought up by their grandparents pressed into them habits of thought, language, and social expectations drawn from the nineteenth century. The Hawkins siblings were fun loving, but formality and propriety were never absent.

When these young ladies took over the management of their ranch, none except Lizzie had yet married. In their new house, on March 5, 1915, two years before the ranch takeover, Lizzie, escorted on the arm of her Uncle Henry, was married to Michael J. Murphy. She was exquisite in her bridal dress of white tulle over pale green satin. Small pink rose buds were embroidered into the fabric. Lizzie's hair was still in the Edwardian upsweep. She carried a bouquet of American Beauty roses. To make room for the wedding ceremony, which took place in the parlor of their new home, Lizzie's sisters had removed the usual furniture, leaving only the piano, and they had made an embankment of ferns and palms against which the bridal couple and the Reverend John Sloan would stand—the same Mr. Sloan who had taught the Hawkinses and accompanied Rowland in applying to the University of Texas. At the piano their friend Marguerite Hamilton (later Gaines) played the wedding march. Lizzie's sisters had filled the house with pink and white carnations, sweet peas, ferns, and palms. They covered the round dining room table with a colorful embroidered table cloth and set in the center of it a crystal epergne filled with mints. Following the ceremony, they served supper to the bride and groom and guests.[1]

In the weeks before the marriage, family and friends entertained the couple with suppers and card parties. Mr. and Mrs. Henry Rugeley had a small supper for the couple and their immediate family. The groom's parents, living in Terre Haute, Indiana, were apparently not able to come to the wedding. Later, in 1924, Mike's mother did come for a visit, and Sister, as yet unmarried and still living at home with Janie and Harry, entertained her with an elaborate party of seven tables of bridge.

The marriage lasted fourteen years but ended in divorce on October 14, 1929, and Lizzie asked the court to restore her maiden name. She continued to live in the big two-story Dutch-roofed house in town where she and Mike had lived. They had no children. The most revealing characterizations of Mike came from my uncle Esker McDonald and my father, neither of whom ever took satisfaction in finding fault. Esker said, "I don't think Mike was very fond of work." My father said he believed Mike overestimated the worth of Lizzie's assets, and a family friend thought Mike liked to gamble at the Cotton Hotel in Houston. These reports hinted at the painful course of Lizzie's life after Mike had entered and then left it. Thereafter Lizzie would conduct her life with a fierce independence and would spiral downward mentally and financially in the years that lay ahead.

On October 15, 1917, my mother and father, Meta Hawkins and James Claire Lewis, were married in St. Mark's Episcopal Church by the Reverend John Sloan. The service was lovely but simple. The bride wore a brown suit, and the couple entered the church together from the transept. From her Aunt Jenny the bride received the seed pearl pin that J. B. Hawkins had given Jenny for her own marriage to E. G. Brodie. (I wore it, too.) After a reception at the Hawkins home in town, a friend, Harold Carter, drove them to the Bay City train station, called "the Depot," for a trip to New Orleans.[2]

Their courtship had clearly quickened in March 1917. Jim, a young Bay City banker, had typed out a letter that he mailed to Meta while she was visiting a Chevy Chase classmate in Joliet, Illinois. He asked for her picture, which she had made in Joliet during her visit; it hangs now in the Hawkins Ranch House bedroom that bears her name. He mentioned sitting for his own photograph. "One of them looks like mud, honest, and I have relegated it to the rubbish heap already. The other one has a sweet seraphic smile and I will send it to you tomorrow. I hope the Lord will forgive that pig headed picture man for all the unkind thoughts he has inspired me with, for I am sure I never will."[3]

The "seraphic" picture is probably one taken in Jim's World War I uniform. He served as a recruiting officer in Matagorda County and surrounding areas, because the authorities thought him too thin to qualify for overseas service. He had to use his own car and buy his own gasoline to cover his territory in his recruiting responsibilities. He was to be reimbursed when he mustered out. Of that time, early in their marriage, he used to tell of two experiences.

On a cold evening, he huddled by a glowing wood stove to stay warm as he worked over the paper receipts and forms that would justify the amount of his government reimbursement. Typically, he decided to cut right through the bureaucratic thicket, put an end to the paperwork, and simply throw away his chance for reimbursement.

> I looked at all that pile of paper, and then I looked at that hot wood stove and those glowing coals, and I just picked up all those forms and papers and stuffed 'em in that stove and burnt up the whole kit and caboodle. Then in a few days I had to go into Houston to report to the officer in charge to be mustered out. I was in a hurry when I was putting on my uniform and, by George, if I didn't get my leather puttees buckled on the wrong leg and appeared before my commanding officer like that![4]

Once the young Hawkinses took over their ranch, Janie was busy every day and made daily drives over the seventeen miles of often muddy roads to check with the hands on the state of the cattle, fencing, windmills, and calf crop. Losing Lizzie and then Meta to marriage and having them move out of the house was saddening to Janie, as Sister told me years later. Janie's sisters were still very close by, not more than a few blocks away, but her household had changed. Their voices did not call down from upstairs, and there were no busy social plans. The tempo had changed and slowed.

Janie had several suitors, among whom was E. B. Wells, who later married my father's sister Letitia (Letty) Lewis. Janie did not so much choose not to marry as she chose to have her life continue as it was. She, Harry, and Sister had made their home together in their red-tile-roofed house in town since 1913 when Sister was seventeen. Janie and Harry lived there from 1913 until his death in 1951; then Janie lived on alone in the big house until her death on February 6, 1958.

Sister was in no hurry to marry. She thoroughly enjoyed keeping house with Janie and Harry, and this arrangement continued until she made up her mind to marry Esker McDonald on September 2, 1927, at South Main Baptist Church in Houston, after a courtship of eight years.

Esker McDonald had become her suitor the day he looked out of the window of Judge W. S. Holman's law office in Bay City, his first place of work after finishing up at Cumberland Law School. What caught his gaze were the four Hawkins sisters passing by in an open car. They all wore big hats; Janie was at the wheel, wearing the leather gauntlets that driving seemed then to require. Their big open touring car, with the spare tire at the side, was decorated in every conceivable place with colorful paper flowers. They were strung through the spokes of the wheels and across the hood and along the doors. Was it the Fourth of July? Returning home as parade participants, they looked like so many bonbons in a candy box. Esker, a little despairingly, said to Judge Holman, "I sure wish I could introduce myself," to which Judge Holman's bolstering reply was, "But you can, you can, of course you can. You are a fine young man."[5]

Sister was pretty, innocently flirtatious, and feminine. She felt herself perennially young and attractive and was so well domiciled with Janie and Harry that it was difficult for her to imagine greater contentment. She had views and satisfactions like those of Jane Austen's Emma, who saw no handicap in spinsterhood except among those whose home and livelihood were less well assured than hers. Sister, living in the home she already had, could plan the menu, entertain with card parties, and oversee Adolf the gardener in stringing up the sweet peas and planting the bulbs. While all her married life was full of love, it was not with headlong passion that on September 2, 1927, she said yes to Esker—yes, she would agree to drive over to Houston with him and Lizzie and Mike and, after arriving there, might decide to marry; but if not, she would ride home again.[6]

Chapter 14
LIZZIE

Those who knew Lizzie in her twenties and thirties, before her troubles began, remembered a beautiful young woman with gloriously golden hair and an air of confident sociability. During the 1930s and early 1940s she loved giving parties for young people—for my brother Frank and his high school friends and for my cousin, Jane Doubek. She played the piano and had collected stacks of sheet music of the popular songs of the day that she kept in a special walnut cabinet with shelves cut to the size of the sheets. A birthday, a graduation, or any national holiday could set her to arranging a card party, dance, hayride, or luncheon. Like Sister she had a feminine willfulness in the way she insisted on giving hospitality. "Have just one more cup of cocoa! You really must!"

The downturn in this charming person's life began on March 31, 1921, when she and Mike had been married only six years. The date was just four years after the young Hawkinses had taken over their ranch from their Uncle Henry's trusteeship, a time when all five were in partnership and fully in charge. On that date Lizzie went to her sisters and asked to have her own undivided interest in the Hawkins Ranch land partitioned to her as her separate property so that she and Mike could operate it themselves. To Lizzie's sisters it was an unexpected and unwelcome move; they expected to be in partnership for the rest of their lives. The circumstance of their childhood had made their safety seem to depend on their staying together.

It is possible that Mike put Lizzie up to asking for her property division. Given Esker McDonald's indication that Mike was not overly "fond of work," perhaps he needed to be occupied with some ranch property to look after. But Lizzie was headstrong, and it is not likely that he could have persuaded her to do something she did not freely choose herself. In any case, contemporary family members never made a villain of Mike, nor did they ever mention any explanation for the divorce.

Lizzie's sisters felt they had to agree to the land separation she asked for, and they even allowed her to make her own selection of the property to assign to herself. She chose a big belt of land out of the middle of the ranch, with farm land on its eastern boundary at Liveoak Creek and grazing land on its western end. Its northern boundary was the south fence of the Sheppard Mott. At the time Farm Road 521, which now approximates that division, did not exist.[1] Lizzie's portion amounted to 4,537.6 acres of land surface and 100 percent of the mineral interest beneath those acres. She designed a cattle brand that she called the Wine Glass, which was a modification of the Hawkins Ranch H-Crook.

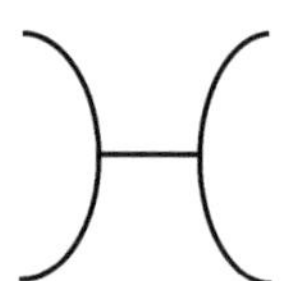

Elizabeth Hawkins's Wine Glass brand.

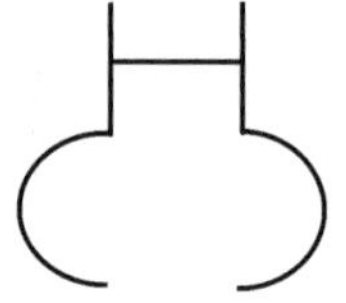

Hawkins Ranch H-Crook.

Although they went along with her wishes, to Lizzie's sisters the partition came as a shocking breach of a kind of unspoken vow they had jointly made long before: they would always stay together and support one another. In the Hawkins Ranch the children of Frank Hawkins had in common the significant place that united them. Now Lizzie had split into two pieces their commonly held land, so cherished because it came to them as a whole from their common past. Even though in the beginning Lizzie's sisters felt no personal hostility in the property division, their regret would deepen into profound sadness in the next decades of Lizzie's life.

In 1935 when Lizzie's sisters and their brother Harry began discussing the condition of the Hawkins Ranch House, Lizzie was not a participant because her ranch property was now separate from theirs and had been for many years. The long-term consequences of this partition were terrible. Year by year, from the 1920s to the 1950s, the fact of the partition slowly built a wall between Lizzie and her family. In the end this dazzling young woman became distrustful and reclusive. Willfully and with destructive pride, she began withdrawing from her sisters' affection, so that finally in self-imposed isolation, she lapsed into mental illness and financial collapse. Then her sisters had to find a means of rescue.

This ruinous decline was slow and at first hardly noticeable; no one thought to put into play any defensive moves against it—not that Lizzie's sisters had the least idea of what the curative moves could possibly be. Their worry about Lizzie was a matter they constantly discussed for decades without ways to do something about it.

At first, once Lizzie was divorced and on her own, she happily and energetically undertook the management of her farm and pasture land. Dressed in riding breeches and boots, she drove to her place every day. She farmed cotton on shares with hands who lived on the place and was exacting in her demands that they keep the rows chopped clear of weeds and well poisoned against the boll weevil. Lizzie had a confident taste for command in farming and ranching as well as socially. "You are lucky," said my mother, "if you can hire someone who has worked for Lizzie first. She gets a person used to meeting an exacting standard."[2]

George Tanner worked as a driver at the Rugeley Motor Company and was a distinguished looking black man much in demand to pinch hit as a caterer when someone was having a party. One day, about 1935, I rode along in the passenger seat as he drove the car to run a few errands for my mother. He got out to take in a package to Lurline Wadsworth, a member of their bridge foursome, and left me in the car with the engine running. I thought: *Now is my chance to make that roaring noise that you always hear when a driver steps on that long pedal on the floor.* I tried it and got the satisfying roar, but George Tanner caught me at it as he returned: "Now what would have happened to you if that car hadn't been in neutral!" My infraction prompted George to tell me about his own need for correction under Aunt Elizabeth. He laughed about one of her punishments when he was a little boy and doing jobs for her, like picking up twigs and leaves in the yard, vacuuming the carpet, and washing windows.

"Ooh, Miss Elizabeth! She was hard on us and made us do right. One time she heard me say a word I didn't have no business to say, and she had me to go find a bar of soap and a toothbrush and wash my mouth out with that soap." He shook his head with conviction. "She learnt me my lesson, all right." Those who worked for Lizzie knew there was a standard to be met and that if there were any breakdown in that standard, there would be consequences.

Lizzie continued to live in town, where she gave her parties, but she created a little ranch house on her place by pulling together several unused tenant farm houses. She fenced in the yard, put in a cattle guard, and planted colorful hollyhocks on each side of it and along the front of the house. The place had large oak trees, too. As a boy, my brother Frank and his friends Sis and Bud Sprouse rode on hot summer days from the Hawkins Ranch House to Aunt Elizabeth's on their slow-going cow ponies, which had to be kicked at every step. Lizzie had Odessa Gatson (later Brown) make lemonade and cookies for them when they arrived. Odessa was then a young girl who attended the county school near Lizzie's place and worked after school for Lizzie; in later years she cooked both for Sister and Meta. "Miss Elizabeth used to come pick me up at school and we'd make cookies for Mr. Frank and his friends. That started me learning how to cook," she said.

Mr. and Mrs. John Ashcraft and their family of lively children lived down the Liveoak Road from Aunt Elizabeth's place. One of the Ashcrafts' sons, Wade, was glad enough as a school boy to pick up any odd job he could at Aunt Elizabeth's. Wade remembered the time when "Miss Elizabeth had us trying to make biscuits with butter instead of Crisco. And we had a time keeping them from falling to pieces because they were so short. But she just kept us at it until we finally got the right combination."

My own memory of Aunt Elizabeth begins about 1935 and in town, when I was sometimes included in one of her entertainments. She had the furniture and rugs taken out of her house for a dance on Washington's Birthday when my brother was in high school. From Houston she hired an orchestra costumed in powdered wigs, silk hose, and knee breeches. My important role that evening, as a grade school girl, was to stand with my older cousin Jane Doubek at the front door and hand a party favor to each arrival. The favor was a cardboard replica of George Washington's "little hatchet," and in each hollow handle was a nice handkerchief. I was thrilled with my responsibility and the chance to watch the dancers.

For her own party dress Aunt Elizabeth had made a trip to the seamstress and specified a dress of filmy pink, with many tiers of ruffles. Ruffles were Lizzie's signature style even though she was now forty-five and not a slender young girl. The afternoon of the party she posed in this dress, standing beside a woven basket filled with gladioli, where a local news photographer took her picture (not included in this book).

Every year I was a real guest at a birthday luncheon on January 25 honoring my cousin Jane Doubek. One year Aunt Elizabeth sent one of her ranch hands and a truck into Houston for a special dessert. Making a big effort for a slight result was characteristic of her. The dessert was pink ice cream molded in individual servings that looked like camellias. The ranch hand and truck did not return from Houston for a long time, and we children kept looking out of the window for it. At last he drove up in a big flatbed truck with a ridiculously small box as his only load. He had made a roundtrip to Houston to fetch the special ice cream stored in a box of dry ice. Hurriedly, the cook found the porcelain dishes for the ice cream camellias, but they were frozen so hard, and our school lunch period was growing so short, that we had to leave our rock-hard ice cream uneaten and dash across the street in time for the end-of-lunch bell.

As an aide in social activities and later at her farm and ranch, Lizzie hired a young woman named Maude, just out of high school, as a personal assistant and driver. Maude picked up the mail at the post office, went to fetch Coca-Colas in ice-filled glasses for guests who needed a welcome in Lizzie's living room, and on most afternoons accompanied Lizzie to the country to check on the cotton crop and the new spring calves. At Christmas time, Maude drove the rounds of Lizzie's sisters to deliver their Christmas presents.

When left to her own preference, Maude wore tailored shirt-waist dresses of fading fabric and awkward length. Little by little, Lizzie imposed her own taste on Maude and, with the help of her seamstress, had her fitted with colorful clothes and, most strikingly against type, a romantic garden hat with flowers at the brim. Maude had been to the post office wearing it, and the bridge foursome took notice. "Meta, have you seen the hat that Lizzie's got Maude wearing? It is quite a transformation. I wonder how Maude likes it." They saw it as just another instance of Lizzie's force in imposing her own will on the hapless Maude. "That's just like Lizzie," Meta laughed.

Year by year, as Maude continued in Lizzie's employ, she seemed to make herself indispensable in the activities of both town and country. Day by day, during the years from the 1930s through the 1940s, many of which returned little or no income from farming and ranching, an imbalance of power gradually developed between Maude and Lizzie. Driving Lizzie to the country, Maude, behind the steering wheel, began to assume the air of the one who did the overseeing. It was she who rolled down the car window to talk to one of the hands, and she who received his hat-tipping deference or mumbled excuses. There came a day when Lizzie said to Maude: "You go on down today without me; I don't feel like going." The slow decline in Lizzie's energy and authority was now coupled with the slow rise of Maude's. Lizzie slipped self-indulgently into ease, and Maude, without Lizzie's objection, assumed pride of office.

I am reasonably sure that Maude did not intend to usurp her employer's whole authority but rather was imitating Lizzie's characteristic hauteur in discharging it. I do not believe that orders ever came from the other direction, from Maude to Lizzie. Maude must have felt that whatever her employer wanted done was serious business and that the more effectively she put Miss Elizabeth's orders into effect, the more elevated her own status became. To Lizzie's family it appeared that Maude was beginning to find ways to lord it over everyone, including members of Lizzie's family, and the situation led to bitter resentment in the sisters. After a few years Maude made it clear to Lizzie's family, when they stopped by to see her, that Lizzie did not wish to see them and that if they came, they would not be admitted.

Sister was closest to Lizzie in age, sympathy, and shared interests. Longer than any of the other sisters, she could still manage to get in to see Lizzie. But the day would come when even Sister was turned away. She rang the doorbell at Lizzie's, and when Maude came to answer and saw Sister, she fastened the latch on the screen door without a greeting or an explanation. Crushed with hurt and disappointment, Sister reported this sad occurrence to the rest of the family, whose impotent response was only shared sadness. Janie occasionally passed Maude, driving alone, on the road to the ranch. There was no friendly wave of greeting from Maude, only icy dismissal.

Lizzie's withdrawal was extraordinarily painful to her sisters, but what could they possibly do about it? That was the question I overheard them asking themselves many times over a period of years. They had no answer. Certainly Lizzie's sisters made a target of Maude in assigning blame. Hers was the face and hand that forbade entry when they went to Lizzie's door. But of course Maude must have taken her directive from Lizzie herself. Why would Lizzie have walled herself off from her family?

A possible psychological explanation was that Lizzie had no capacity to admit failure. Like an actress, she played the role of success and avoided financial reality. By every tactic available, including extravagance, she denied what was happening. She could manage the shame of failure only by projecting an enemy who had caused it, and judging from her behavior, she apparently came to regard her own family as that enemy.

Lizzie's clear directives, ordinarily a virtue in management, were in her case a terrible disadvantage. She tended to act on intuition and impulse and was not given to analysis. Lizzie was not the first person who has sought a livelihood in agriculture whose eye fell less on the ledger than on the condition of the grass, crops, and herd. Sometimes it is said that the best fertilizer for a ranch is the footstep of the rancher: a saying of dubious value. Everywhere an owner looks there are things that need to be fed, doctored, mended, planted, drained, mown, or bulldozed, and all of them cost money. The farmer-rancher cannot know what the revenue is likely to be until long after the expenditure, when the product is sold at the current market price, whatever that may chance to be.

Lizzie was operating during the Depression years, which were hard for everyone, including her sisters, but she could not shrink her flamboyant style. My family and I happened to drive by her house one day and saw that her garage door was open. Inside was a new Cadillac that had been backed into the garage so that we could see its beautiful hood ornament and elegant black curving fenders. "My goodness," my parents gasped.

The day came when Lizzie failed to pay her bills—could not pay them; when she sat at her desk, she kept shoving aside the unopened window envelopes from Houston's big stores, Sakowitz and Battelstein's. They lay piling up in the pigeonholes of her desk, as was discovered after her death.[3]

Lizzie's recourse was to borrow money. With the help of Victor Le-Tulle, a successful farmer and banker, she put up her land as collateral and borrowed enough to sustain her ranching and farming operation for a while. When she had new shortfalls, someone advised that she might sell off a small percentage of her mineral interest. As noted, when her land was partitioned to her in 1921, she had received 4,537.6 acres and all the mineral interest under it. Selling part of that mineral interest would not interfere with the grazing and farming of the surface. There was not enough mineral exploration at the time to suggest that selling some percentage of it was giving up much value. This advice probably came from Will Davant, a highly regarded lawyer in town and a trusted friend and advisor to Lizzie. He had the habit of stashing his bulky legal documents in the inside pocket of his suit coat, making it list to one side as if a flatiron were hidden there. My father used to say, "If Will Davant ever lost his coat, everybody in town would have to go out of business."

On Sunday, December 8, 1935, my family and I happened to drive by the First National Bank and were surprised to see a group of people entering the side door of the bank on a day when the bank was ordinarily closed. My father said to Esker, "My goodness, what is going on? Look at that gathering of people at the bank and on a Sunday!" Both being bank officers, they were doubly surprised to see a meeting of some kind. We caught sight of my Aunt Elizabeth. Then my parents, having had some forewarning, realized that she was there to arrange the sale of part of her mineral interest and that Mr. LeTulle, president of the bank, had opened it on the weekend as a courtesy to my aunt. The day was cold, and Aunt Elizabeth, wearing a full-length mink coat and a leopard-skin turban, was in the company of Will Davant, Victor LeTulle, and several other gentlemen, who were the buyers of her mineral interest. She stood with them for a moment just outside the side door of the bank, waiting for the door to be unlocked before they all filed into LeTulle's office. This transaction did not end Lizzie's troubles; it only bought her some time.

Davant probably felt he was coming to the rescue of a lady in distress. Here was an interesting, attractive woman who was relying on his advice. Once when I was in the car with my parents, Aunt Sister, and Uncle Esker, taking a Sunday afternoon drive through the Hawkins Ranch, we decided on the spur of the moment to skirt the Currie place, cross the Liveoak Creek bridge, and bump along through the Four Mile Bottom toward Hawkinsville, the site of the old Caney plantation. It was a route I loved because it passed through an absolute jungle that always seemed mysteriously full of startled animals. Sitting in the back seat between my aunt and uncle, I caught a grasshopper that sprang in through the open car window. I pinned its sharply angled hind legs between my thumb and forefinger and held it for a while before letting it make its spectacular arching leap back to freedom. My father said, "I wonder if he knows how to get home, now that we've gone way past it." A tender little worry invaded my thoughts.

Just as we entered the woods from the open prairie, I reached out of the window to catch hold of a bright green sienna bean and then opened the pod to expose the neat row of beans. There was always one bean on the end much smaller than the rest. "That's like the runt of the litter," my father said. Sienna beans grew as tall as the car, and as we drove, they whipped against the fenders of the black Ford. After a ride like this, when we got home, we always found long brown wilted stems caught in the crevices. Weeds and brush like this were a menace to the pasture land, especially near the woods where they seemed to thrive. Cutting them back with a mowing machine pulled by a tractor was ongoing work, often interrupted by mechanical failure.

The trail through the woods was only wide enough for one car, and the driver had to watch carefully for brush, fallen limbs, or hidden stumps overgrown with vines. Several times my father, who was driving, had to get out and walk ahead a few paces to see what obstruction or low muddy spots might lie just ahead. I loved this mysterious route because it transported me right into the Tarzan movies with all the cawing birds and buzzing insects. As one entered the dark woods, out of the leafy damp scent, there arrived a kind of presence. I begged my father to take our family someday to Africa, where I imagined this sense of mystery was perpetual. "Oh when could we go?" His answer crushed me because he named a time that seemed impossibly far off. "When you are about nine," he said.

Continuing through the woods and rolling over small branches that snapped under the tires, we heard another car approaching from the opposite direction. We were all puzzled to be meeting anyone on this hidden trail known only to a few people. The car came into view. Both cars had to pull to a stop to figure out how best to pass while also avoiding the low spots off the road. Out of the car came a smiling Mr. Davant, and with him in the passenger seat was Lizzie. They greeted us, smiled, and waved. The passage was then negotiated, and we went on, and they went on. Once we had driven a few hundred yards, Esker said, "Well! What do you think of that?"

Sometime during World War II Lizzie's creditors were ready to foreclose on her farm and ranch property. None of her siblings was able to approach Lizzie directly; every friendly initiative they took was rebuffed, and harshly. Lizzie's sisters and brother decided to see if they could pay off all her debts and, in effect, buy from her creditors the property that had been partitioned to her in 1921.

On November 6, 1945, the arrangement was made through intermediaries, and the portion of the Hawkins Ranch that was Lizzie's Wine Glass Ranch came back into the Hawkins Ranch, making it whole again. The mineral interest Lizzie had sold did not come back, but the estimated 50 percent interest she still held did rejoin the Hawkins Ranch holdings.

Lizzie's family was never able to surmount the sadness they felt that even though the Hawkins Ranch land was now back in one piece, Lizzie herself remained lost to them. They had to ask themselves how she would be able to support herself.

They decided to provide her with a fixed monthly income from the Hawkins Ranch. They wanted to be sure she would actually get the check and have the money to use herself, and they knew from all their attempts to see Lizzie that she would not accept the check from them directly. They called on Wade Ashcraft, who had by then become a banker and, as a young bachelor, was living in an apartment above Lizzie's garage. She had known him and his family since he was a boy, and he was a reliable, comforting friend to her. He became a courier, delivering into her hands the monthly check or such cash as she needed from it. Through him Lizzie's sisters and brother could have news of her well-being. At least Lizzie's family had lifted from her the bitter discouragement of a failure that her pride prevented her from facing.

Late in Lizzie's life, her family hugged to themselves one last hope that Lizzie might come back to them. In 1957 Wade called to say she was ill and needed to see a doctor. She had become a Christian Scientist and had also simply been heedless of her own health needs. Her siblings saw this new vulnerability as a potential window for them to reach her again and bring her back into the family fold. Her illness puzzled the local doctors, who advised taking her to St. Luke's Hospital in Houston. Some combined heart and kidney problem was the diagnosis, and she stayed in the hospital in Houston for a few weeks. Her sisters brought her flowers, cologne, and hand lotion in a crystal bottle with the emblem of Imperial Russia on its side. She and Meta clasped each other's similar, slender fingers. Sister plumped up her pillow, smoothed her bed, and gave her a pink bed jacket. Lizzie seemed to enjoy their attention. Perhaps she would come back to them. Things seemed to be better. She was alert but weak and willing to talk to members of the family. The doctors were optimistic. Then, with no warning, she died suddenly on March 3, 1957. The cause of death, as determined by an autopsy, was a ruptured artery near a kidney. She was sixty-four.

Chapter 15

THE CONVERSATIONS IN THE FAMILY, 1935

From childhood, I remember that Lizzie's troubles were discussed with more worry by the Hawkins family than any other issue, because of being so painful, so long in duration, and so beyond their control. But another issue preoccupied them in 1935. Should the Hawkins siblings let the Ranch House—the Lake House, where they were born—fall all the way down, as had the Currie house, or should they try to get hold of someone to repair it? Unlike their father and grandfather, who had managed the Hawkins Ranch while living on the place, the Hawkins sisters ran it while living in town. They did not need a headquarters on the ranch itself, and the house had fallen into neglect and disrepair.

The question now on their minds was whether to let nature take the house all the way to its extinction. It would be a natural transition, familiar to anyone who went into the country and viewed old houses after their proud purpose had come to an end. People of the Gulf Coast were used to seeing the process of destruction and considered it a part of life itself. Driving out into the country, one could always see the remnants of some deserted dwelling located in a once convenient place at the corner of a field now unused. The winds of seasonal northers would have torn at a corner shingle and exposed the house's interior to rain until nails and planking loosened and the structure began to lean and finally fall. The process took time to happen, but inexorably it did happen. If time and change had divested a structure of all its practical purpose, then its only remaining purpose must be an attachment to the story the structure might tell, but was the story now worth the lumber and the nails?

The Hawkins family returned many times to the question of the Ranch House, rather than settling the issue all at once. As a child I overheard snippets of these conversations, and through the years I got to know the leisured way the family decided things, especially worrisome things. In conversation, they would try out a new idea or course of action; only eventually would they set off in one imagined direction or another until they felt the worry beginning to ebb. If they shared a feeling of comfort that rose up like an unsought breeze, they would know they had all decided. Their ways were never brisk; they would never have called a meeting in order to decide; efficiency they made a lower rung in the order of being. Then, too, the family's sense of time always suggested to them that there was plenty of it. Often their deliberations took place at leisure times when they were together anyway. They enjoyed spending time together, because they were friends as well as kin, and they had been brought up to be good company.

Some of the conversation about the Ranch House took place at Matagorda Beach on Sunday afternoons just before sunset, while the adults sat around on camp stools bending over paper plates and cautioning the children about the undertow. "You see over there where the river runs into the Gulf and the water changes to that reddish color? That current will just pull you right out to sea." The river was the Colorado, which rises north of Austin and makes a diagonal slash across the middle of Texas through Austin, Columbus, Wharton, and Bay City and finally empties south of Matagorda into the Gulf of Mexico.

Some of the conversation probably took place on the grounds of the Ranch House itself. Around 1935, on weekend outings, we sat around a fire built in the yard rather than going into the house to make a fire in one of the many fireplaces, because the corner of the house had fallen off its supporting sill, heaving up the dining room floor and setting it at a steep angle, unsafe to cross. The house was not safe to enter, to say nothing of the danger to anyone venturing to climb the winding staircase to the second and third stories. The condition of the house was something we could hardly ignore on such an outing.

The Currie house, a favorite Hawkins family picnic spot, served as a reminder of what the Hawkins Ranch House was likely to become. Its ruin was within a dense wood west of Liveoak Creek, about five miles to the east of the Hawkins Ranch House. The Currie house had belonged to an earlier Texas family named Cavanah, who settled there during the colonization by Stephen F. Austen in the 1820s. For many years the Hawkins siblings had watched the disintegration of the house in that tender way one watches an old friend grow older. It was still standing in my childhood, but when we picnicked nearby, we children were cautioned not to play too close to the listing brick chimney. The house and the more than seven hundred acres around it did not at that time belong to the Hawkins family; it was the Curries who, in the absence of decision, let it fall.[1]

The land on which the Currie house was located did not so much adjoin the Hawkins Ranch as cut into it, in a narrow horizontal thrust two thirds of the way through the middle, as if a fruit knife had been left in the side of an apple. This penetration, coming from the east at Liveoak Creek, ended barely short of the Hawkins western boundary at the shore of Lake Austin. For many years the Hawkins family leased this land, because without it they would be confined to a narrow passageway through which to move cattle between south and north pastures. Since 2002 the Hawkins Ranch partnership has owned the Currie property.

The Currie house had been allowed to collapse even though it had a more compelling story than many houses in Texas. The sentiment to preserve it did not outweigh the inutility of doing so. The historically significant dog trot design of the Currie house was still apparent. It was built of two boxlike wings made separate by the dog trot, an open corridor running from front to back. A second story loft space was built above the first story and served as a cover for the open corridor or dog trot between the boxes. In 1935 the remains of this house were all but hidden by tall pecan trees and clumpy yaupon that grew thick along Liveoak Creek. There was a large wild plum tree there, too, which excited my mother's interest. She was always receptive to the folks from the country who brought her small tubs of wild plums for making beautiful deep red-amber wild plum jelly, into which she never added a trace of pectin.

In approaching the Currie house we left the wide open prairie and entered thick woodland, where a small clearing and a large fallen tree near the house was the picnic spot. My mother would bring out the fried chicken and stuffed eggs and begin a search for the key to open the can of potted ham. I and any childhood friends I brought along would be sent to gather twigs to start a fire. "Oh, you can't start with limbs that big," my father would say. "You've got to begin with a bundle of tiny little dry twigs no bigger than matchsticks and then add pencil-sized ones, you see, and after the fire gets going, you can go to those bigger ones you have there. You've got to give it time, or you'll smother it." My father was very much at home dealing with practical matters of country life or any matter that entailed a process and an identifiable outcome he could plainly see as beneficial. He could have been the father in *Swiss Family Robinson.*

Much of the ground around the Currie house had become overgrown with brush, but there was still enough clear space for picnickers, and a few stumps and fallen logs provided seating. For children it felt haunted, because the adults were bound to retell the story of the Karankawa Indian massacre of 1826:

"You know what a 'massacre' is? It means everybody in the house, the whole Flowers family, were killed, just wiped out. All, that is, except one little girl."

"Whew!"

"And the reason she was not killed was that those Indians thought she was already dead."

"Maybe she played possum," we said.

"She had an arrow stuck in her back and those Indians gave her up for dead and left. Then she got up and escaped to the Van Dorn family on Liveoak Creek, nearby, and they took care of her. She was the only one to be saved. Fourteen people were killed."

The fact of the massacre is undisputed, but there are variations in the details. My mother's version was that the lone survivor was "a baby left sleeping in a crib and found later by the Van Dorn family." Aunt Janie's version was that there were two wounded girls thrown into a brush pile and left for dead but that both recovered and escaped. The massacre is described in *Historic Matagorda County* and other sources.[2]

If the collapse of the Currie house were to be their example, the Hawkins family would simply let their Ranch House fall back to nature. In the tug of war between practicality and sentiment, where would the family place itself? Practicality favored allowing its natural decay. Such an antebellum house was no longer a compelling necessity in 1935, when ranch management could be handled from town, and the 1854 design of the Ranch House presupposed a way of life that simply no longer existed. Broad galleries needed to be swept and mopped. Chamber pots were emptied and washed daily. Mosquito nets needed to be fixed to the bedposts at night and taken down in the morning. Wood had to be chopped to fit the size of the cook stove. There was no electricity or indoor plumbing. The design of this house and the way the design suggested living in it were not the way anyone would have built a ranch headquarters house in 1935. Neither did any of the current members of the Hawkins family plan to live there. Even if they restored the house, they would never live in it again; they had established homes in town. How strong were the strings of sentiment that pulled them back to the place?

Chapter 16

JANIE AND HARRY

While the Hawkins family often spent Sunday afternoons together at some county beach or wooded spot on the ranch itself, they were also likely to gather on a week day in the evening on the screened porch at Janie's house in Bay City. The time of day chosen would have been "after supper," in the early evening. Meal times for all households were scheduled, sit-down affairs, and visitors courteously delayed an unannounced arrival until after supper. Gatherings like this were also likely settings where the condition of the Ranch House was discussed.

The porch at Janie's was next to the dining room, and on evenings when the family gathered on that porch, Janie and Harry would have finished their supper. Typically it would be a thin steak, grits, tomatoes with cucumbers, biscuits, and for dessert, Jell-O with heavy cream. After the meal they would put away their large white dinner napkins, folding them and pushing them through their respective silver napkin rings. Janie's was an oval one with "Janie" engraved in cursive script; Harry's was octagonal with his initials in block letters, H.B.H. for Henry Boyd Hawkins. I was often there too, a child who had stayed too long and been invited to supper. If I had been in the way, the family's code of hospitality would never have let that fact be noticed.

Our family's meals had some elements of performance and could be times of sociability, to which all seated at the table, whether children or adults, were presumed to know how to contribute. To be at the table was to make an implicit pledge to maintain pleasantness by steering the table talk toward engaging topics. Adults repeated anecdotes and consulted one another about forgotten kinships: "Was he the one that Cousin Emma married just before the 1909 storm when they lost all their furniture?" Meals were about enjoying well-prepared food but also about knowing how to respond to social cues or how to be inventive when none was evident. Table sociability did require squandering big chunks of time three times a day at a place where one counted on good company and felt an obligation to be good company oneself.

Aunt Janie's dining room opened onto the south screened porch through French doors, each pair with floor to ceiling squares of beveled glass that cast little rainbows in the evening sun. The porch was furnished with green wicker rockers and a wicker porch swing, hung from the ceiling by lengths of chain. In summer Aunt Janie kept the doors open, and the breeze made chimes of the prisms hanging from the big upturned bowl of the awkward Edwardian chandelier above the round oak dining table. At the end of a meal, it took only a gesture from Aunt Janie, as subtle as the start of a string quartet, to suggest taking a seat out on the porch. Children understood, more by observing than by being told, that these rituals were not for the sake of formality but to make pleasantness continue. That pleasantness was already in being, and as a general condition should be made to continue, was a central assumption, the assumption of habitual politeness.

Social convention was not an onerous duty to them; it was a matrix through which social inventiveness, which is to say charm, found its chance. Comic breaches of social convention, like those of W. C. Fields or Laurel and Hardy, the family found hilarious. Tedium they did not in the least dread, because they were practiced at touching up the ordinary and drab with their fairy dust, the effect of which was laughter.

Harry, the eldest of the five Hawkins children and the only son, had been lame from childhood, yet every day he walked, slightly scraping the toe of his left shoe, the few blocks into town to get a shave and, more frequently than necessary, a haircut. The trip to the barber's was his first order of the day after his breakfast of shirred eggs, bacon, and Holland Rusk toast. He needed a daily shave because, having a weak hand, he was not able to do the job himself. Mainly, however, it was a pleasant walk to town, with a greeting by the barber and his other customers and a comforting set of services. When Harry finished at the barber's, he stopped in the confectionary called the Alcove for a "cold drink." Sometimes he treated Miss Iris at the jewelry store to a drink too. He was fragrant in the mornings with shaving lotion and hair tonic and in the afternoons with cigar smoke. He had little else to do. If he were asked the time, he drew out his large gold pocket watch and opened the case carefully but awkwardly. Speaking slowly, he would pronounce, "It is fifteen minutes until nine." He never said "eight forty-five" or "a quarter to." I never heard any reason given for his lameness or passivity. A moment's lack of oxygen at birth? His disability was part of him; his family, enjoying him just as he was, had no reason to trace out the cause.

Sitting on the porch, enjoying the evening's bird sounds and puffs from his cigar, he was as still as Buddha. He had distinguished features, a Franklin Roosevelt profile, and he liked having his picture taken. His brothers-in-law, my father Jim and Esker, teased him for being so silent and distinguished; they called him "Judge"—"Now, Judge, we've got to have those birthday barbecues a little more often. You are just not letting those wheels turn fast enough!" He was flattered by family teasing and attention and shared the family's ironic sense of humor. After the noon meal every day, he would say, "Now, I am going to my office," as he climbed the stairs to take an afternoon nap in his bedroom.

Placid and opinionless though he was, he could laugh heartily at radio comedy. Let King Fish and Andy start up about the Mystic Knights of the Sea, and Harry's high-pitched laughter would carry all the way to the dining room from the sitting room, where his chair was pulled close to the brown wooden console. Aunt Sister enjoyed telling of the time when she and her sisters took Harry with them to the opera in Houston, the results of which made them double over with laughter when the events were recounted. "When the tenor jumped out on the stage and let out the first note, Harry fell into such a laughing fit we all were asked to leave. They just threw us all right out!" If he drank a bottle of beer in the late afternoon, as he often did, he would acquire an amused expression and recite a radio jingle: "B double E R, it picks you up and never lets you down." I was the likely audience for this little recitation.

One morning when I arrived at Aunt Janie's by tricycle, a trip of two blocks from my parents' house, he encountered me on the sidewalk as he was returning from the barber's. "Good morning, Miss Margaret; it is charming to see you!" He bowed as if I were a queen, and if he had had a good arm and a plumed hat, he might have swept the air in one of those wind-up court bows like Cyrano's. He was being playful and teasing. Sometimes he gave my playmates and me a beautiful gold metallic paper ring from one of his cigars. And when I needed a box for my crayons, I could have a pungent wooden one on the lid of which was a picture of a beautiful lady in a white blouse with her dark hair drawn up in a high Spanish comb.

Harry's sisters honored him every August 21 with a birthday barbecue on the grounds of the Ranch House. After a few years a screened wooden shed known as the Barbecue House was put up permanently just for such occasions. At first it was nothing but a shelter from the sun and had a dirt floor and large screened openings on the sides, without glass panes. It was meant only to provide a shady place to cut a watermelon and serve the mutton barbecue and Aunt Sister's potato salad. The Barbecue House, always associated with Uncle Harry's birthday, was rebuilt several times over the years. Its later improvements included a floor, a fireplace for winter heating, and a built-in barbecue grill with an exhaust fan.

These birthday barbecues in the Ranch House yard brought together our extended family and cousins of various ages. We younger children chased around and collected fireflies in jars, and the older boys of my brother's age caught the big calves that happened to be in the lot and tried to ride them. They let out great war whoops and inevitably got bucked off. My cousin Martha Rugeley (daughter of Rowland) and her friends were a little older than I and pleaded to try driving the family car or at least to put it into gear and ease around the yard. I felt very flattered if I were asked to ride along as a passenger in their company. When we finished, we usually went for a crawl through the bales of hay stored in the barn.

Barbecuing required that a ranch hand select a young sheep out of the herd and string it from the bois d'arc tree to butcher it. A wood fire had to be built up and allowed to die down. Making a good barbecue sop was my mother's job. She used butter, vinegar, Lea and Perrins sauce, ketchup, lemon, and a dash of sugar—all boiled down in a little water. Aunt Sister's contribution, potato salad, came in a huge mixing bowl of golden, smoothly blended mashed potatoes mixed with onions, pickles, salt, pepper, mustard, and, I am pretty sure, a large amount of butter.

About twenty members of the family and friends sat at long unpainted wooden tables with Uncle Harry at the head. With fanfare, a birthday cake with lighted candles was brought out. Usually it was a gold or chocolate cake with caramel icing and chopped pecans in the filling between the layers. Children were happy to oblige Uncle Harry in the job of blowing out the candles and singing "Happy Birthday." The ice cream we had was a doubtful success. On the way to the ranch for the party we stopped at Stinnett's, the second of the two confectionaries on the square. Will Stinnett then packed a gallon container of chocolate ice cream and a gallon of vanilla and placed them in a thickly padded brown cloth case, stacking one container on top of the other. This giant sausage he tied tightly with the attached cloth tabs. Half of the contents always melted, but floating in creamy liquid remained two large lumps of chocolate and vanilla available for serving.

The family gave Harry presents in colorful wrapping paper and bows, and we children helped with the unwrapping. There was a round crystal humidor with a metal top designed to look like the Sphinx; a dictionary-sized black leather case with a snap closure for carrying a dozen cigars in the automobile; a tulip-shaped silver cup engraved "Harry Hawkins," just right for a dram of hard liquor, which he never happened to take. It taxed the birthday guests to think of presents for Uncle Harry; his needs were so few. I don't think he was ever given an item of clothing; one of his arms was shorter than the other, and Aunt Janie had all his shirts and nightshirts made at the Hamilton Shirt Company in Houston. He had no need for a billfold; he carried his bills and coins in one of the stout, double-sided leather snap-shut purses that the bank gave out.

Uncle Harry would sit on the porch in dignified silence while he smoked his cigar. If asked his opinion, he nodded in agreement with the opinion expressed by the others. Why would he not? Everyone around him was full of affection and good intentions toward him. What his sisters and their spouses thought to be a good course of action had always seemed so to him, too. His comfort and contentment bore that out. About the Ranch House and its condition: well, whatever the others wanted to do about it was fine with him.

One afternoon, coming home to my parents' house from one of my rambles down the alley, I entered through the kitchen of our house as usual. My mother and her bridge foursome were taking their break for coffee and servings of butterscotch pie. The foursome sat at a bridge table permanently set up in the very small library of our house. I think it was Daughty, Rowland's wife, who said, "You know Meta, isn't it amazing how much Janie looks like Amelia Earhart?" And all the others chimed in to agree.

In the mid-1930s Amelia Earhart's exploits as an "aviatrix" had been pictured in the newspapers and newsreels. She was heralded for flying across the continental United States and was not yet the tragic figure who disappeared in 1937 somewhere in the Pacific. Pictures showed her ready to climb into the cockpit, a smiling young woman in breeches and boots standing beside her airplane, the wind from the airfield tossing her short hair attractively. The bridge players expressed admiration for the "wind-blown" hairstyle, as it was called. Amelia Earhart was not only a pilot; she was also something of a fashion designer. My mother had acquired an "Amelia Earhart jacket" of golden pigskin that she said was soft and comfortable for car trips to the ranch in chilly weather. It had a zipper at the front and four zippered pockets—a design Amelia Earhart sponsored.

Janie at this stage was a youthful forty-three, and she had the same wind-blown short hairstyle and appealing smile. In spirit as well as appearance, there was indeed something of Amelia Earhart in Janie; being out of doors and active was her way. The country life interested her far more than the domestic one.

Like her sisters, she was fair with reddish hair and dressed carefully for the indicated occasion. When she went to church or to a daytime function, she wore a carefully selected suit and a pretty hat with a brim. For the daily trip to the ranch, with Harry in the back seat of the Chevrolet and holding the strap, she wore riding breeches, western or English boots, and a long-sleeved shirt of cotton, wool, or satin—depending on the season. Always she had a colorful scarf around her neck fastened with a pretty gold equestrian pin. She pulled the scarf up to cover her nose and mouth if the cattle were being herded that day and were kicking up more dust than usual.

On country outings Janie carried in her ample drawstring purse more than one pair of pigskin gloves, which she rolled up like socks, and—in case she had to open a wire gap in the fence—a pair of pliers.

"I wish you'd look here in Janie's purse!" a family friend named Georgia one day exclaimed, on discovering the pliers.[1] "I would just like to ask you, have you ever in your life known anybody in the world but Janie who would carry a pair of pliers in her purse?"

"That's all right," Janie said, "but you all know who's got a pair of pliers when you need 'em. And you just watch, you're going to need 'em."

Somewhere in her car Janie also kept a long flashlight requiring many batteries laid end to end; it was like those that the police of that day might carry in search of a perpetrator. And in winter, when the air at the ranch was likely to be colder than in town, she took along not just one extra coat but two, "because," she would say, laughing, "if somebody needs a coat and doesn't have one, I don't want to have to give up mine."

If cattle had to be worked for shipping, she had a horse saddled and ready when she arrived at the ranch, and along with the cowhands, she cut calves out of the herd. When I was of high school age I joined her in that. Cutting calves out of the herd puts the horse and rider in happier unison than usual, because cow ponies, though needing to be urged when on the open road, were stimulated by this activity. Quickly they sensed which calf one intended to select, and they became animated, almost prancing in anticipation of each darting turn of the calf.

In quail or dove season, all the family went to the country prepared to take a shot at a bird or, if need be, a rattlesnake. Janie carried a twenty-gauge shotgun, exactly fitted to her arm and shoulder with a red rubber recoil pad glued to the butt of the stock. A protective leather sheath was lashed onto the gunstock by leather thongs. She had discussed the fitting thoroughly with the local gunsmith and locksmith, Mr. George Helmecke, and I had been taken along when she had some service done to the gun at his shop. Mr. Helmecke was known among other things for continuing to drive a Model-T Ford long after the Model-A supplanted it. His business slogan on a wooden sign out in front of his shop read: "We make every kind of key except Whiskey."

People said of Janie that she was "one of a kind" and had a "dry wit." The bridge foursome whooped at what she had to say about a particularly pleasant but rather homely young woman in town, for whom everyone wished and hoped the best, although doubtful that such hopes would be fulfilled. "Well," said Janie, "I have looked at Pearly Hughes [not her real name] from every angle, and nothing can be done"—hope advanced and struck down all in one blow. The foursome appreciated too the way she summed up Mr. Engel, the Episcopal minister, whose sermons Janie and most of the foursome had heard many times: "The trouble with Mr. Engel," she said, "is that he is simply overcome with the joy of living."

At parties with many tables of bridge, Janie was known to kibitz and appear to give authoritative advice, not knowing a thing about the refinements of the game. "You should save that one," she would whisper, "to catch back in." She would laugh delightedly when found out.

On her trips to the ranch Janie equipped herself to deal with unexpected inconveniences or minor accidents. Ointments, dressings, liniments, and doses of patent medicines of all kinds were a hobby with her, and she was usually well supplied and ready to treat anyone she could persuade to admit to a complaint. Absorbine Junior, an antiseptic, was a special favorite, and when the Phillips people took milk of magnesia out of its exclusively blue bottle presentation and offered it in small white pill form, she heralded the breakthrough to anyone not yet in the know. "Try one of these," she would say, holding out a small tin box with the blue printing of the Phillips Company.

Her grandfather Rugeley's house, where Janie and her siblings grew up, was the home of a doctor, and she remembered how "Grandpa Rugeley" compounded his remedies out of a black leather case neatly fitted with glass vials of liquids and powders. At the turn of the century when he was in practice and Janie was about ten, the doctor's role must have seemed authoritative and glamorous. Certainly, in that day and in their town, a doctor was accorded high social standing. Dr. Rugeley's wife was inevitably referred to as "Mrs. Dr. Rugeley" and sometimes as "Mrs. Dr." Perhaps as a young person, in some unconscious recess, Janie had had an aspiration to be a doctor herself, but the idea, if there, would hardly have been a tenable one in her time and circumstance.

When the conversation in the family came around to the condition of the Ranch House, Janie's reaction was not the expected one. Although she took an active interest in the family cattle ranch, she shrugged off the Ranch House as a lost cause and was ready to let it fall into ruins. Years after the fact, my mother told me of Janie's indifference, which we agreed seemed odd. Perhaps her reluctance stemmed from the expense involved and from the fact that, as a single woman, she did not have a husband's resources to add to her own. Yet she and her brother Harry made their home together and, sharing their household expense, they managed well. I think the real reason for her reluctance was that she was temperamentally disinclined to change anything at all or to take a new action of any kind unless it was unavoidable. She loved the out of doors and was in that sense activist, but that did not mean she was inclined to undertake and oversee a major house renovation. She took no pleasure in redecorating, and her initial decision about the Ranch House—her default position—had already been taken. It was simply to leave things be.

That way of thinking, I imagine, is what Janie's intuitions and blind spots consisted of. I do know that from 1913 until her death in 1958, she lived on in the house in town where all the Hawkins siblings had made their home in their young single years. Even when she had full charge of the place, she never once changed the location of a single piece of furniture or agreed to repaint or repaper unless some ruinous damage had occurred.

Chapter 17

Sister and Esker

When Sister and Esker arrived at any family gathering, it almost always created a little laughter, a little stir of delight that everyone felt when they joined the group. Esker's charm derived from a capable, take-charge generosity. Sister's came from a captivating vulnerability and the ease with which she laughed at herself. If there had been some news of the day that came from her encounters with the grocer or the hairdresser or the cook, she would create some laughter out of it at her own expense.

My parents, Jim and Meta, often walked over to Janie's on summer evenings from their house down the street, as did Rowland and Daughty, my great-uncle and aunt, who lived within the same block. Sister and Esker rarely walked. There was only one time that I can think of, and it came in later years, that Sister would have been willing to walk three or four blocks: when she was trying to take exercise in order to "reduce." For the few weeks in their lives while she was undertaking this effort, Sister and Uncle Esker got out after supper and walked a planned route together, as if he had weight to lose too.

Esker was tall, manly, and gifted as an athlete. He was also a splendid ballroom dancer who knew how to give his partner a confident lead and to keep perfect time with good grace. He taught me the Charleston. As a boy he had played baseball and had briefly been a practice pitcher for a semi-professional team. "I learned a lot that way," he said. In school he played football, and early in his marriage he took up tennis. It was played at night on a lighted court made of concrete. The local editor and sports writer, Cary Smith Jr., praised him for a feat of sportsmanship when, instead of taking advantage of his opponent who had slipped down, he chose to return an easy lob to give his opponent more time. Later, when a golf course was built, he was the course champion year after year.

My father and Esker were officers in the local bank. As the cashier of the bank, Esker signed his name with model handwriting, finishing with a little flourish, just for the pleasure of the motion. He was quick to hold a chair for a lady, to open the door for guests and give them a warm greeting, and when departing guests needed help backing out of the driveway, he was out of doors in a flash, waving them on.

He was a favorite with children because he was patient and enjoyed instructing us in practical skills we found enjoyable, like teaching a dog to fetch. He knew some tricks with coins and handkerchiefs, and he had strong arms and shoulders and could lift us up with an easy swing. One evening in my childhood when I had supper with Aunt Sister and Uncle Esker, he announced for my benefit that as supper was finished, he would now challenge himself to peel a whole apple without allowing the peel to break at any point.

"Everybody watching? Here we go." Starting at the top of the apple, he created a very long red ribbon that curled down toward his knees. He and I were delighted with the achievement.

Sister was five feet, two inches tall, plump with a pretty face, porcelain skin, and blonde hair. In riding to the ranch with Esker driving, she held up against the passenger window a folded newspaper to keep the afternoon sun from tanning her face. She and Esker had no children and, as a consequence, made children of each other. Sometimes she went to the piano, and he and she sang (slightly off key) something they had heard from a Jeanette McDonald and Nelson Eddy movie. Sometimes it was "Ah Sweet Mystery of Life."[1] But sometimes, with her diamond rings flashing, Aunt Sister ragged out "Five Foot Two, Eyes of Blue," and he would say it was an exact description of her. He treated her like a cherished, fragile toy. And she was feminine and nurturing without limit. When he had a cold, she clucked and murmured at his bedside, plumped up his pillow, smoothed out his sheet, put a cool cloth on his forehead, and brought him tea, soup, or cocoa.

"Oh law," she announced one day, when he was sick, which was not often. "Dr. Livengood said it's a wonder I hadn't killed Esker. Dr. Livengood told me to make him some good homemade soup, and I had all that sweet corn and heavy cream; so I fixed up some corn soup that was good and rich, and he ate every bit of it and really seemed to enjoy it. Then when Dr. Livengood asked me what kind of soup I had given him, he said it was the seventh wonder I hadn't killed him."

She loved a rocking chair and was never happier than when rocking someone's baby to sleep. "Sister can knock 'em out quicker than anybody," my father said. Certainly she loved comfort and loved nothing better than giving it. But she often judged what was needed in terms of what she herself preferred. "My goodness," a friend of mine said of her, "when I stayed overnight with her that time, and she came at me with that down comforter, and it was already eighty degrees, I just had to stand my ground."

Rowland, in Sister's presence, shook his head in mock disbelief and pretended to sum up her character: "Sister is the only person in creation who can lead a horse to water *and* make him drink." She did have a feminine determination when she had a project and could require a great deal of the cook or the yard man or the store clerk, in the same way that Shirley Temple had of stamping her little foot. She could be especially forceful in giving hospitality. "Now you come on, you must have just one more piece." In the afternoon the reference would have been to the ever-present box of Russell Stover chocolates freely offered to children, whose mild, guilt-ridden protest she would completely override.

Somehow, because she was an innocent, and a loving, generous innocent, she unwittingly cast a spell upon everyone. Her charm and vulnerability planted in everyone the sense that, whatever the inconvenience to oneself, she by all means ought to be accommodated. If she were not, the sin of omission was like that of failing to stroke a sweet kitten or take the hand of a small child. Her evident vulnerability gave her unsought power; it meant that to oppose her was all but unthinkable because it required that good people turn downright mean.

In some magical way, too, she had a salutary effect on children to whom she made her insistent appeals to "just have one more piece, go right ahead!" Somehow, the extravagance induced in them a kind of inward decision to take charge of themselves. Some interior voice would say, "This is a little crazy. If I am really being urged to have all my wants indulged, then I had better make some judgment of my own about the advice I am being given." Magically, her very insistence let judgment arrive by some route and edge out total self-indulgence. It was an effect like that aspired to (whether fulfilled or not) in the educational philosophy of A. S. Neill of the now-defunct English school Summerhill, where the policy was to remove all rules, to permit almost total freedom, so that children might begin to take charge of their own impulses rather than suffer the pain of unbridled chaos.

As the youngest of the five Hawkins siblings, Sister had no memory of childhood days at the Ranch House.[2] Harry, Janie, and Meta did remember playing childhood games of twelve sticks and a cattle rustling game played with corn cobs representing cows penned up in the owner's pen. Cattle pens were marked off by sticks placed end to end in a circle, and the object of the game seemed to be to slip past surveillance and steal a neighbor's cows.

Even without these childhood memories, Sister was more enthusiastic than any of her siblings about saving the Ranch House from decay. I can be sure of that because I saw her in action on other fix-up jobs, sitting in a rocking chair, her place of command, keeping a close watch on Johnny the painter and spurring him on. She loved to have a project and was as enthusiastic for this one as Janie was reluctant. Any project was exciting to her, especially one that involved choosing colors, wallpaper, drapery fabric, new sheets and towels, or soaps and bath salts with lovely lavender scents. Prettiness and making things pretty were her special delight. Her taste, however, ran to eighteenth-century porcelain and gold leaf, which she would confidently have thought just the right touch to enliven Thoreau's drab cabin that he was so backward as to leave unadorned. And she loved the color pink so much that she put it in every possible place. One year she had a pink Christmas tree. "Pink just makes me feel good," she said.

Sister could be an enthusiast, too, because she so entirely discounted the difficulty of any endeavor. Her naïveté about practicalities allowed her to push forward without deigning to hear any objections or discouragements. And she could push a carpenter or painter hard. Casting her spell over him, she would magically waft away his inertia and objections. Faced with her sweet unreasonableness, he would begin to smile. He would wag his head and emit a patronizing chuckle. He imagined that he condescended to her, so obvious did it seem to him that he knew and she did not. He had no idea that he had been charmed by her effective vulnerability. So he went to the work she insisted upon, sure that she needed his rescue.

Chapter 18

META AND JIM

My mother Meta, the eldest Hawkins daughter, was highly in favor of proceeding with renovation of the Ranch House and impatient to get on with the project. If Sister was propelled by dismissing the practicalities, Meta was propelled by the consideration of them. She thought dithering too long was to lose an opportunity. "It's just going to wrack and ruin if we don't do something about it, and we ought to get started on it." Impatient impulse was her way. "Just do something, even if it's wrong," she would say.

Meta had the same fair complexion as her sisters but redder hair. People teased her as a child and pretended to light a match from her hair. Her skin was fair, but her eyes were brown. She was the only one of the Hawkins siblings who had children. When she was thirty years old my brother Frank was born, named for her father, Frank Hawkins. When she was thirty-eight I was born, delivered at home by Dr. Livengood. Another child, a little boy, born between Frank and me, died at the age of two. His smiling baby picture in a wooden frame was always on her dresser. He was called J. C. for James Claire, my father's name. Fifty years after J. C.'s death, I happened to ask my father about the cause of death. Exercising control of his emotions, he said, "I always laid it to a cow we had." J. C. had infant dysentery and died of dehydration and shock in 1927.

Meta described herself as a jack of all trades and master of none. She had amazingly nimble, long slender fingers and enjoyed working up Chopin etudes, which she never quite worked up to finished form. Her piano practice and the lessons she sometimes took as an adult were not toward the end of playing the waltzes and swing pieces her bridge partners Lurline Wadsworth and Irby Stinnett often played to provide impromptu dance music.[1] Nor did she seem to favor the sentimental tunes that Sister played.

Meta's playing was more for the purpose of reengaging a discipline she had enjoyed while a piano student of Philip Tronitz at Kidd-Key College in Sherman.

I had some oil paints in tubes that our artist friend Georgia Mason Huston had given me, and to my surprise one day my mother took them up and painted a beautiful full-blown rose on a green tin laundry hamper. I had no idea that she knew how to do such a thing. She must have learned the trick in taking china painting, which all the sisters had taken up at one time. In this or that cabinet we had creamy yellow pitchers, teacups, and small trays each covered in flowers and having one of the Hawkins sisters' signatures: Meta, Janie, or Lizzie. Of the pieces with Janie's signature Meta said, "I really doubt that Janie ever finished hers; I think the teacher probably just finished it for her."

After breakfast every morning Meta made her progress through the yard, stopping to uproot a weed or take the head off a dead flower. The yard man was ever present and under her direction. She had no hesitancy in giving firm directions or, when she thought necessary, an equally firm reproach if there had been some misjudgment by which a young plant had been taken for a weed and uprooted.

Routinely, the breakfast table, the luncheon table, and the Sunday dinner table had flowers from her yard. Sometimes the lowly zinnia was all; at other times she arranged camellias and roses. She had a kumquat plant in the side yard and was always "outdone" when the school children stripped it clean of its orange thumb-sized fruit.

"Meta sets a good table," was said of her by other women who set good tables themselves. She knew what good food should taste like. She knew how thin and in what diameter hot cakes should be served—the size of a teacup and with plenty of egg and not too much flour in the batter. Biscuits should "have no insides." She was unshrinking if she needed to point out to the cook that the biscuits had been rolled too thick and had too much baking powder, that the pie crust was a bit tough this time, or that the meringue had grown thin and watery. But at the end of dinner I also heard her say to the cook, "You have simply outdone yourself!" And my father would say, "You know what they say the Mona Lisa is smiling about? She's thinking of a piece of banana cake." A bright smile would then beam out from the kitchen.

On the afternoons when the bridge foursome was not in session, my parents often went into the country when my father came home from the bank, to check on things and see whether rain had damaged the corn crop; to talk to Rose and Douglas Chapman about the seed corn, poisoned against weevils, and how careful they needed to be about not letting any animals get hold of it; or to check on why the sow had not had her shoats.[2] My mother and father were wonderfully companionable on these country sorties. They both understood and enjoyed the management of land and the details of farming. For my father especially, to ride into the country was recreational. It stirred up boyhood memories of rambling along Wilson Creek and shooting some game or catching a fish for his mother.

If anyone were to do anything about the Ranch House, my father Jim was the person most likely to round up someone to get the work done. As a small town banker he talked every day to farmers and carpenters, druggists and car dealers, and, when spring came, to school teachers needing a loan for summer study. He always had good information about people who had special skills and might want work. Although he was wary of intruding, he strongly agreed with my mother's opinion about the Ranch House.

"A house like that, built, let me see, before the Civil War, that's a rare thing in the county and, the truth is, in the whole state. Why, if the Currie family hadn't moved off their place, if there had been any of their people around here to see about it, they would still have that dog trot house standing right today. As it is, all they have is a pile of rotted-out wood and two chimneys falling into brick bats. You don't want that to happen to you. And besides, it's the place where you all were born." That last observation, reverently made, signaled that to lose the house would be to lose a significant part of the heritage in the family to which his wife and children belonged, a family that held his loyalty.

Chapter 19

Rowland and Daughty

On evenings when Rowland and Daughty joined the family on the porch at Janie's, they had only to walk up the alley a half block to come through Janie's back yard and enter the porch through its screen door. They lived in the house of Rowland's late father, Dr. Rugeley, where the Hawkins children had grown up.

Even though Rowland was like a brother to the Hawkins children, he would have had no proprietary responsibility in deciding what to do about the Ranch House. It was in his character to be scrupulous in avoiding intrusion into decisions that were the Hawkins children's to make. If he thought it helpful, he might offer information based on his own experience with refurbishing projects. He knew the carpenter C. K. Norcross and considered him a good man for a building project.[1] But it would also be like Rowland to sound a caution: getting into a job like that could be a journey with no end. There was no way in the world anyone could know what the cost would be. The price of lumber and shingles could skyrocket in the middle of the job. Norcross was fine, but they had better watch to see what kind of helpers he got and whether they would stick with the job and stay sober, because he could not do a job like that all by himself, although he was a loner and liked to work that way. If Rowland said something cautionary of this kind, I suspect that my mother and Aunt Sister would both have pounced with something like, "Oh pshaw, Rowland. If you think too long, you think wrong!"

The Hawkins children appreciated Rowland, who had Janie's same dry humor, but they all thought him maddeningly precise. His wife Daughty teased him about being so thorough in his morning shave that he made use of two razors. He would laugh and say, "Well, you know I put a lot of reliance on those Old Enders, but two are better than one." He carried in his pocket a pair of small fold-up scissors in a leather case, on the off-chance of discovering a loose thread at his shirt sleeve or buttonhole.

Rowland was a handsome man with sandy hair and a spare physique. He was always dressed with bandbox neatness, often sported a bow tie, and wore a small gray Stetson hat. For the benefit of the bridge foursome, Daughty would affect wonderment at his success in doing business with the local dry cleaner, whose service was a frequent topic for them: "I don't know how in the world it can be that Verser Brothers just puts everyone else's clothes through the ringer of ruin, but Rowland somehow can get them to turn his out perfectly crisp and clean."

Daughty regaled the foursome with how Rowland stayed ahead of her in kitchen tag. "I was getting ready to make a chocolate cake and got out a pound of butter to soften on the counter top. Then here comes Rowland, sees it, and of course thinks someone had carelessly left it out, so he puts it back in the box. Then, I come in again and see that it is gone, and there it is in the box! So I put it out again. And, bless pat, if he doesn't put it straight back in the box the second time! I tell you, you just can't get ahead of him!"

After graduating in law at the University of Texas in 1912, Rowland had taken a job as an assistant to County Attorney Dick Lewis before his attention was diverted from the law. When automobiles first came into use, town officials had to put up signs advising drivers to "keep to the right side of the street" rather than crisscrossing or circling as if riding a horse or driving a wagon. In those early days Rowland began to sell a few cars and eventually built up a Chevrolet and Buick agency.

Janie and Harry stopped by his garage for a fill-up before their trips to the ranch, and he would emerge from his office for a chat through the driver's window while one of his garage men pumped a lever back and forth, to make the pink gasoline rise to an etched mark on a big glass cylinder at the top of the pump. Then the pink would slowly edge down the cylinder as the gas drained into the car's tank. It was a favorite sight of mine when I went with Aunt Janie and Uncle Harry to the ranch.

Rowland's early legal training and, to a greater extent, an inherent character trait disposed him to be exacting in testing any truth claim. One had to be careful about making any sort of assertion, or there would be questions. "How do you know that? What is the evidence for that? Are you really sure of your facts there now?" These questions were asked with a kind of merriment, as if he were teasing, but he was not. Somewhere along his educational path, he had developed such a respect for the factual underpinnings of statements that he would not allow a sweeping generalization to go unchallenged.

Rocking in the green wicker porch chairs at Janie's as twilight came, her family could switch on the ceiling lights, but June bugs and moths would then storm the lights. Often they switched off the lights to sit in the soft darkness, with little more on their minds than being together and enjoying the respite from summer's daytime heat. No doubt they sometimes discussed the Ranch House.

One way of thinking about it was to let nature make the decision for them. The destruction nature wrought would be so gradual that the family would scarcely notice their own complicity in choosing to let it happen. Maybe they should just let the house crumble away and die a respectable death. That course was the least expensive option and the least troublesome, because it appeared to bypass the need to decide. As mentioned, Janie was comfortable with that tactic. No one had lived in the house for decades, and no one in the family intended to live there now.

On the other hand, the house was their birthplace, and it stood in the middle of the cattle ranch that their father and grandfather had built up piece by piece. Ranching was their livelihood, but it was also more: ranching on this land fastened the family to their past, and this ranch house was central to that past. Sister and Meta and my father Jim took this view of the matter and leaned toward rescuing the house.

When the Hawkins family were together, wherever they might be—at the beach, at a picnic in the woods, or visiting on the porch at Janie's—they did not intentionally raise the question of the condition of the Ranch House. Their conversation would simply have drifted to it. I am sure the family did not make a decision about the house until they had talked it over many times in many settings, because I do know what made them decide. It was the coming of a lady visitor and what her visit revealed to the family about the house.

Chapter 20

The Lady Visitor and the Decision

The family's decision about the Hawkins Ranch House came one day in 1935 after the arrival in town of a lady visitor, Vera Prasilove, and her young daughter Kytja, who was about my age. I was pleased to be asked to come to Aunt Janie's during the day to be company for Kytja. As the younger of only two children in the whole family, I was adept at observing things, unnoticed; but I did not know how to be useful. Up to that day, the only time I remember being in the least helpful was when I was asked to hold my index finger against a ribbon to assist in tying a Christmas bow on a gift probably destined for me. But now I was actually being sent for and seemed to be needed.

Right away, though, I began to doubt that I would be of much use in providing hospitality. The exotic name of the daughter and mother placed the pair beyond my social experience, and the little girl's saying "porridge" instead of "oatmeal," and "cockroach" instead of just "roach," revealed that she was a stranger to this area of the country. I was made timid by the awareness that we were each at the edge of the other's region of understanding.

Mrs. Prasilove was Czech, and her husband, Dr. Arthur F. Scott, was a member of the chemistry faculty at Houston's Rice University, then called Rice Institute. By far the most arresting thing about her was that she had a profession: she was a photographer. In my young experience, although men had professions, there was hardly ever an occupational name associated with women. Nearly every woman identified herself in terms of her husband's occupation: "My husband has a dry cleaning business." Or "My husband is the postmaster . . . the druggist . . . a rice buyer."[1]

In a small town there was hardly ever a need for either men or women to provide identification because everyone knew them anyway. And children, it was assumed, were fully summed up when an adult said, "Whose little girl are you?" I doubted my ability to create a welcoming bridge for Kytja, because I had had no practice whatever in introducing myself to a stranger or making someone feel welcome. Everyone I met in my rounds of the town was someone I already knew and who already knew me.

Fay Guynn, a family friend from Houston, had arranged for Vera Prasilove's visit, thinking that the Ranch House would be a good subject for a photographic study. I can look up as I write this and see the silver gel photograph she took of the house (not shown in the book). She signed the picture the way a painter signs a canvas and dated it 1935. And so I know how old I was that day when I stood beside her as she set up her heavy brown wooden camera box on its wooden tripod and put into the camera a square that looked like a pane of glass. I was seven.

To take the picture, there first had to be an automobile trip through the country to the Ranch House. Aunt Janie led the way in her Chevrolet, with Uncle Harry as a passenger, holding to the strap. Kytja and I and all the photographic equipment followed with Mrs. Prasilove, in her car. I had only seen my aunts take pictures with a small Kodak box camera or one that had black accordion-pleated bellows stretching out in front. Mrs. Prasilove's camera did have the pleated bellows but was itself at least a foot square and was made like a big polished wooden box. Jutting out from the front was the camera lens with a black metal surround. The big box was awkward and heavy, and what was more, an adjustable wooden tripod was also needed. And then there was a black case for holding glass photographic plates and other pieces of equipment.

Before the county had made improvements in roads generally, driving to the ranch entailed passing along a street east of the town onto a road that was hard-surfaced only on one side, with the unpaved half left roughly topped with oyster-shell gravel. Those conditions meant that when Aunt Janie's car met another, either she or the oncoming car had to vacate the paved side until the two cars had passed each other. We arrived at an unpainted wooden gate leading off the Liveoak Road, to turn onto the Hawkins Ranch land at its northernmost pasture, the Sheppard Mott.[2] Upon our arrival at the gate, the African American children who lived along the Liveoak Road ran out to do us the favor of opening the gate, which was not locked but merely fastened by a chain looped over a bolt in the gatepost. They hoped for a quarter for this little service and, when the transaction ended, exchanged big grins with us in the car.

Then came a dusty ride along a private road through the Sheppard Mott. We passed creeks and sloughs and counted the cranes standing with one foot raised. Field larks flew up, jackrabbits dashed out, and sometimes a slow-moving armadillo groped its way along. Small black birds we called rice birds flew over our heads, moving in a swirling ribbon as if some inborn command said, "Choreograph! Then children will see you and lift a hand to the sky."

As we passed out of the Sheppard Mott heading south toward the house, we saw the clean straight line of the horizon separating the blue above it from the green below, a line almost entirely uninterrupted by any silhouetted structure except distant windmills and a few clumps of huisache. The clouds spread themselves out, and we could see immediately whether they were billowing white cumulus or those arranged in horizontal streaks of salmon color. The sky of the coastal plain is vast and captures the attention of anyone who drives into the country; clouds were closely observed to learn if a change in the weather was likely. Despite the vastness of the sky, the horizon seems very near. And the earth looks flat. I always felt we could walk or ride a horse over to the edge of the earth itself. What empirical evidence could overturn what lay before our very eyes?

Soon we began to catch a hazy view, not of the Ranch House itself but of what from a distance looked like a gray mound that pushed up the horizon into a low arch. The arch was made by the grove of oak trees Ariella had planted in front of the house so many years before. When we arrived at the wooden front gate of the house yard, Mrs. Prasilove set up her camera.

"Now I will play hide and seek," she said laughing. I was too literal minded to understand that she only meant she would now duck under the black cloth thrown over the camera. She stayed with her head hidden for a minute or two. "Would you like to see?" What I saw under the black cloth was a shock. The familiar symmetry of the house—upside down! A little disgusted, I came out and looked straight at the house itself, with its pairs of square columns ranging along the upstairs and downstairs front porches. The narrow glass panes—lights, they were called—on each side of the front door glinted softly in the afternoon sun. When the house and these glass panels were later painted by our artist friend Georgia Mason Huston, she used for the panes of glass a little purple and moss green, because their reflection was never the blinding, white kind that a mirror would throw back.

It was afternoon as we stood there with the camera, viewing the front of the house. The sun threw light in gold parallel strips across the pathway leading to the wide front galleries. The arched leaves and branches of Ariella's oak trees entirely obscured the three tall narrow dormer windows jutting out of the roof. Nor could we see from the ground the widow's walk, a railed platform that sat at the very top of the roof like a square crown. We did not go up to the widow's walk that day. Had we been allowed to climb up to it by the dark, narrow inside stairway on the third floor, it would have given us a panorama of the flat coastal country that had in it no hidden crevices. Its unimpeded horizon was drawn as if with a crayon in one lateral sweep.

From up there we could have seen the road over which we had just come from the north. To the south, we would have seen the upper shoreline of Lake Austin. Sometimes from up there, way in the distance, one could see a wisp of dust rising from the road, and someone would say, "There they are now." There was no telephone or electricity or indoor plumbing in 1935. If visitors were expected, someone watched for them; but if they ran late in arriving, we usually found out only when they did arrive.

The photograph Vera Prasilove made that day in 1935 is still lovely, and when one stands today on the spot from which the picture was taken, one sees the same house, unchanged in design. In the house portrait taken that day, there were no people, just the house itself. But Sister and Janie were each asked to pose for a portrait and did the next day.

Sister's is the farthest removed from her interests, habits, and tastes, which ran to Dresden figurines. Mrs. Prasilove posed her standing on the open prairie. Her gloved right hand holds a horse by the rein near its bit. She is dressed in boots and riding breeches, which were her own. She wore a canvas cloth riding coat she had borrowed from Janie. The cloth belt of the coat is fastened with a large button in front. In her left hand she holds at her knee a soft white cloth hat like a cricket player's. It was Esker's. When the photographer had looked through her lens and declared that the composition would be even better if her subject held a hat in her hand, Esker characteristically provided the remedy: "Take mine," he said.

Aunt Sister's own saddle was on the horse, a simple cow pony and not handsome. I could always identify the saddle that was hers when all the family saddles were hanging in the cistern house; hers was the one with the stitched, comfortably quilted seat. In the picture her face is lighted by the afternoon sun. A shadow becomingly shades the curve beneath her high cheek bones, and a little touch of light puts a blush on the tip of her nose. Her hair is casual but parted. It was soft around the face; the 1930s were far past the time of both the voluminous upsweep and the chopped-off look of the radical bob. Aunt Sister's pose as a young lady rancher, though uncharacteristic, was a pretty one.

But the portrait of Aunt Janie mounted on her horse was decisive in the family's deliberations about the Ranch House. That photograph made all the Hawkins siblings sharply recognize the same circumstance: something *had* to be done about the house! Aunt Janie was dressed in her riding clothes and astride her horse. She had on her riding gloves and wore a soft felt riding hat, a long-sleeved blouse, and a riding vest. She posed close to the south façade of the Ranch House, with a shuttered window and the side of the house close behind her. Painfully, the resulting picture exposed the peeling paint of the house and the loose and missing teeth of its shutters. This house was the birthplace of the five Hawkins children, and it was falling to pieces like an old, unneeded hay barn left to degrade into rot and kindling. The photograph taken by the lady visitor amounted to an embarrassing public exposure. The fixed, still picture served as a catalyst to action. Although the family had often seen the actual peeling paint and toothless shutters, although they knew from actual experience the condition of the house, this photograph of its declining countenance—the publication of its condition—propelled them to repair it. They decided; no wisp of doubt remained. Nor did they doubt that Norcross was the right man for the job. My father would get in touch with him.

Chapter 21

THE RANCH HOUSE AND MR. NORCROSS

Everyone knew that the only way to deal with C. K. Norcross was on his own terms. His workmanship and integrity were so respected that most people needing a job done would wait until he was ready—wait for his crankiness to subside. Sometimes a pint bottle prolonged negotiations until gentleness set in again, and then, people said, that there was not a man around who could do a better job for you.

My father held the secret of good rapport with Mr. Norcross. He sometimes loaned Norcross money to tide him over when carpentry work was scarce or he needed to hire a helper to work on a job. Sometimes my father made the loan out of his own pocket, if the circumstances were not right for an institutional loan.

When the Texas Gulf Sulphur Company abandoned its mine in Old Gulf, near Matagorda, about 1936, it put up for public sale a number of good wood frame company houses no longer needed for employees.[1] My father bought one and had Norcross move it up the Matagorda highway to Bay City to fit it out as rental property that would provide some extra income for my father's widowed sister, Emma Carleton.

"Now, how much do I owe you, Mr. Norcross? You don't want to delay any longer giving me your bill; it's been too long now, and I want to pay you." That question was the start of one of his favorite Norcross stories because the reply was so characteristic of the man. "All he ever said," my father laughed, "was 'Oh, just skip it, skip it!'" And that was the end of that. They both knew that skipping it was a way of repaying a personal loan, but Norcross was not going to lose dignity by going into details.

People thought of him as a crotchety old bachelor, but he had once been married and had a daughter named Juanita, whom I came to know many years later when we both worked in a Houston bank. But something must have gone wrong; he had obviously determined on being single and seemed to take pride in being grouchy or in making himself appear that way. Underlying the crust was vulnerability and good will. He was never sharp with us children even when we surely were in the way, watching him work. His softer side he hid with a scowl, a tightly clenched pipe, a wry sense of humor, and a craftsman's attention to the business at hand. He had thick gray hair and never wore a hat or a cap even in bad weather. The story of his aversion to hats came to me from Miss Tenie. As a boy he had been required to dress nicely and to wear a cap he detested. One day he agreed to wear the cap at his mother's insistence as he set out for school, but once out of her sight, he threw it into the pig pen, where the pigs helped create an unsalvageable mess of it. His daughter Juanita, by contrast, prided herself in budgeting to buy an occasional designer suit and Christian Dior hosiery. She and her father were not estranged; they simply had no needs the other could fulfill.

Norcross set to work to reclaim the 1854 Hawkins Ranch House. He put on his blue bib overalls with a loop at the hip for his hammer and a pocket for his fold-up carpenter's rule. Several pockets at the chest housed a flat wooden pencil for marking lumber, his gold pocket watch and chain, and a tin of Prince Albert pipe tobacco. He hired two helpers and simply moved down to the Ranch House yard, camping there most of the time for more than a year while he worked on the house. As he required lumber and materials, he drove to town in his Model T Ford. Sometimes he made the trip back from town with two long flexible planks bent over the top of his vehicle, the ends tied to the front and back metal bumpers. When he needed large pieces, he hired a bigger truck. One huge piece of lumber that he used to reinforce the interior structure of the house was a second-hand piece, a very long sturdy board originally used, judging from the way it was cut, in a two-story staircase. Some of the original wood in the structure of the house, revealed by the renovation work, had been hand-planed with an adze, and the nails used when it was built had square heads.

One of the first tasks was to get a jack such as a commercial garage would use and attach a long rail for leverage, so that Norcross and his helpers could jack up the sagging corner of the house and level it. He put new sills beneath it and, to make doubly sure it never sagged again, he built a low brick supporting wall running east to west beneath the house. It is there to this day and securely intact.

He finished the interior ceilings with beaver board, a new material at the time, consisting of squares of some brownish rough-surfaced stuff, reconstituted, I suppose, from wood pulp. Beaver board was a serviceable and cheap finish but not handsome. Sheet rock was not a material then in use, so to repair the original plaster walls, a plasterer was brought in to restore the places where mildew had softened the plaster and made it fall off in shards. The deep cornices, some as deep as twenty inches, and the walls throughout the house were of plaster.

In time, electrical conduits were installed in plain view on the interior walls so that some of the rooms could have ceiling lights. At first, in the 1930s, the source of electric power was a series of automobile batteries strung together in what was called a Delco system.[2] The batteries were housed in a small play house I had had in town and were recharged, supposedly, by the power of the windmill. More often than not, however, they were as dead as a door nail. Later a rural line came into being; only then was there a reliable source of power.

Kerosene lamps with graceful glass chimneys had been the main source of light in the late 1930s. There was no indoor plumbing, and none was added by Norcross in the renovation of 1935. A three-holer outhouse on the east side of the front yard was discreetly hidden by trees. For overnight use, a lidded white enamel chamber pot was placed under each of the four-poster beds on the second floor. Because mosquitoes were always a nuisance, Norcross attached screen doors to each bedroom door frame on the second floor. He happened to have in his tool box a heavy brass door latch that he claimed had belonged to Thomas Jefferson and had been used at Monticello. This, for some reason, he installed on the door of the east bedroom that we call Aunt Sister's Room. It is there still. Bedrooms were on the second floor, and the third floor consisted of two unfurnished bedroom-sized rooms and a large loft that had been used in the 1850s and 1860s as a dormitory for itinerant tradesmen. This third floor space was unused throughout the twentieth century and was left unchanged until 2001, when the third floor loft space was reconfigured to create a third bedroom and two bathrooms.

By about 1938 Norcross had made the house safe to enter, and it came alive again for the family. Entering the house by the front door, one faced the typical symmetrical arrangement of a house of the period. The paneled front door had slender vertical windows (called lights) on each side. These were made of rectangular glass panes, each 8 by 18 inches, set vertically side by side to rise to the height of the door. To avoid risk to glass and give support, a wood panel was placed at floor level instead of a glass pane. A rectangular transom window above the door was made of a straight horizontal line of ten vertically placed panes like those in the sidelights. A white wood frame outlined and supported the door, the sidelights, and the transom—visually enlarging the centered entry and making it suitably proportioned to the large three-story house. Just above this ground floor porch entry was another, identical to it, centered at the second story porch.

On entering the front hall at the ground floor, one viewed an expanse of wood flooring made of ash and cut by J. B. Hawkins at his sawmill. The balustrade and handrail of the stair, according to Norcross, were made of mahogany.[3] Up the stairs, which rose straight long the west wall on the right, one could climb to the second story and beyond. A sharp left turn at the top required careful footing as each riser narrowed. One then had access not to a landing but directly to the second story hall. From there the same type of staircase rose along the west wall to the third story with the same sharp turn and narrowing risers. Floor openings, protected by the balustrade continuing from the stair, allowed a view of the ground floor from the third.[4]

In the entry hall of the ground floor, three doors presented themselves. On the left (east) side a door led into a formal room called the parlor. Originally it was furnished with high-backed Victorian settees and matching armchairs, all in dark mahogany with turned legs. A large marble-top table was placed in the center of the room. The furniture was the kind that ladies wearing long skirts and achieving excellent posture found comfortable. It was a room where a tray of hot chocolate and hot, yeasty Sally Lunn could be set down on the marble-top table, as my mother recalled was a custom in her childhood. The parlor was a fine room for festive occasions, but it was never intended to have the warmth and coziness of the room opposite it on the west side, entered by the door on the right as one enters the front hall. This room was called the sitting room. In today's terminology it would be called a den.

If on entering the front hall you went straight ahead through the facing doorway, you entered the very large dining room, where with the use of leaves the table could be made to seat twenty-four people. After the repair of the Ranch House this was the place of large gatherings of cousins and aunts and uncles and friends for Thanksgiving and Christmas. Those usually attending were those who came to Uncle Harry's August birthday barbecues, but in December the season brought us indoors and sometimes made us huddle by the fireplace.

Christmas and New Year's Day menus were inevitably the same, because they could hardly be better. There was turkey cooked in a wood stove because at the time the uncertain electric system did not measure up to supporting an electric range and oven. The cornbread dressing was seasoned with shallots, green peppers, celery, salt, black pepper, and the rich broth from the turkey. No seasoning of sage was ever introduced. Cabbage was a perennial; cooking it a very long time with a thick slice of salt pork made it meltingly tender and surrounded it with delicious pot liquor. Sweet potatoes were an accompaniment, along with rice and wonderfully seasoned gravy. Uncle Rowland always said there could never possibly be too much gravy. Elizabeth Rugeley, one of our cousins, sometimes prepared wilted lettuce salad with coleslaw dressing that always won high praise. Either Uncle Esker or my father would carve the turkey—an exacting, skilled task that was the prerogative of a senior male family member. Wine was served. For dessert, thin small wedges of mince pie were brought in with mounds of whipped cream on top.

Often someone brought a cake. If it were my mother, the cake was my brother Frank's favorite, a gold one with caramel icing and a filling of chopped pecans mixed with the icing. Ambrosia was also a favorite dessert, made of layered slices of orange, pineapple, and flakes of fresh coconut—the kind one could get only by hand-to-hand combat with the coconut itself. The oranges came from the grove of orange trees Ariella had planted in the 1880s, grown from stock given her by her husband's brother, Dr. Alexander Hawkins, who at that time lived in Florida. Those trees bore fruit plentifully and still do. The oranges are juicy but full of seeds, so it took a little work to make them do for ambrosia.

The house was heated entirely by wood fires. Altogether there were nine fireplaces in the house, all shallow from front to back in order to throw the needed heat into the rooms. The fireplace surrounds were made of black cast iron, some of which showed the date of the casting to be 1829, long before the house was built. Every room had cold drafty spots, and people tended to huddle near the fire when there was a brisk norther.

James B. Hawkins reported that the original free-standing kitchen was blown flat in the hurricane of 1854, at the time he was building the house. He did not say whether he replaced this free-standing kitchen after it was destroyed. Perhaps he did. My mother seemed to remember that a separate kitchen had burned. If so, it would have been one that J. B. Hawkins rebuilt after the first one was destroyed by the storm. The original design of the house and dining room indicates that food was first carried from an outside kitchen into the small pantry that still adjoins the dining room on the west wall. From the pantry an outside door leads onto a side porch, which would give access to an outside kitchen.

The only kitchen I remember was a small one that seemed to have been tacked on to the south exterior wall of the dining room. That kitchen worked well enough for many years even though it was nothing more than a passageway from the dining room to the small cistern house. The cistern house was a semi-open porch covered by a roof, with walls made of slats spaced slightly apart. It protected the round brick storage receptacle for rain water used in the household. The kitchen was not changed until it was enlarged and improved in 2001.

Norcross covered the big round brick cistern with a lid of heavy boards, so that it became useful as a low table or seat. The roof line of the cistern house always caused comment from visitors because the east side of its roof was much longer than the west side, giving the roof a startling asymmetry. Norcross offered the opinion that the purpose of the roof over the cistern was to collect water and that, since the prevailing wind and rain came from the east, the down spout leading into the cistern was made to run from the larger side of the roof on the east side. This asymmetry was eliminated in 2001 when a new, larger attached kitchen was created.

The second floor of the Ranch House was an almost identical layer of rooms on top of the ones downstairs. There was a large front door to the second floor porch, directly above the ground floor front door and matching it exactly. Above the parlor was the east bedroom, which came to be called Aunt Sister's Room. Above the sitting room was the bedroom called Aunt Janie's Room. Above the dining room was a small east-west hallway opening at each end onto side porches. A narrow outside service stairway was accessible on the west side porch. Originally firewood, water, and laundered clothes were carried up these stairs.

Crossing this east-west "cross hall" that gave access to the upstairs side porches took one to the bedroom on the south side of the house, the one we called Meta's Room. Each of these bedrooms had a fireplace. Aunt Sister's room had a view of the orange grove on the east. Aunt Janie's had a view of some of the trees in the front yard and of the cattle pens. Meta's south view looked out at Dode's house and the bois d'arc tree beside it, a tree that still stands today; beyond is the shore of Lake Austin, where she remembered her father, Frank Hawkins, had a boat dock for a small steam vessel. The wooden piers were still visible in 1935. Cattle grazed beside the lake.

No wonder the question of the house was so long discussed in the family. In 1935 it was not easy to pay the cost of repairs. The expense was shared by all the Hawkins children except Lizzie. When it came to painting or wallpapering the second floor bedrooms, the Hawkins sisters decided to designate a room for each of them and to bear the cost of decorating individually. The rooms they elected to sponsor still carry their names; Sister's was naturally enlivened by many touches of pink.

The five Hawkins children after their move to town to live with their Rugeley grandparents (collage), ca. 1899 or 1900. Left to right: Elmore (called Sister), Lizzie, Janie, Meta, and Harry. HRLTD

The red-tile-roofed house under construction in 1913. This was the Hawkins children's first home of their own. HRLTD

On the south porch of their new house. Left to right: Janie, seated; Sister at the door; Lizzie behind the trellis; Meta, seated; and Harry. HRLTD

Sister in a rowboat near their house, during one of the frequent floods. For more about floods see Stieghorst, *Bay City and Matagorda County*.

The Hawkins children's uncle Henry Rugeley (left) was their trustee when they lost their parents. Beside him is Sister, and next to her is Rowland Rugeley, Henry's brother. Photograph taken at the wedding of Henry Rugeley's daughter Mary to William B. Ferguson Jr. (called Tilly), who stands behind the bride, March 23, 1922. HRLTD

Janie on horseback among cattle to be shipped. Photograph by William B. (Tilly) Ferguson Jr., 1930s. HRLTD

Janie (right) stoops to examine a calf with the ranch hands and veterinarian, Dr. Marks (bending, with only legs and hips visible). In the foreground at left is Paul Ford, and William Franklin and Dode Green stand at the fence. Photograph by Beryl Sprouse, 1940s. HRLTD

Automobile parade in Bay City. Meta is in the big hat in the car at left. Janie is at the wheel of the car at right, and Lizzie stands on its rear bumper. In the background is the courthouse built in 1896. The sign at left reads “Bay City Realty Co.” The date of the photo is before 1928 when a new courthouse was built; likely about 1917. HRLTD

Jim Lewis asked Meta for her picture in March 1917. She sent this one, taken while visiting a school friend in Joliet, Illinois. Jim and Meta married in 1917. HRLTD

Jim Lewis gave Meta his picture. He served during World War I as a recruiting officer and is pictured in his uniform. HRLTD

Sister posed in 1923, four years before her marriage to Esker McDonald in 1927. The hall chair was acquired at the time the Hawkins children's house was built and furnished in 1913. HRLTD

The Currie house, painted by Forrest Bess. In the 1930s Meta and her sisters commissioned the Matagorda County artist to paint this house portrait. For decades the Hawkins children observed the slow disintegration of the Currie house (also called the Cavanah house). HRLTD

Sister, 1935, photographed by Vera Prasilove. At the photographer's request, Sister holds a hat; it is Esker's. HRLTD

Janie, 1935, photographed by Vera Prasilove—the picture that triggered Ranch House restoration. At left are the smokehouse (near) and the tool house, not major barns. To the far right the tacked-on kitchen is visible. The diagonal line across the photograph is not a wire but probably a flaw in the photographic process. No electric power served the house at the time of the photograph. HRLTD

Part III

The Instruction of Town and Country

Once it was built in 1854, and especially after it was saved from decay in the 1930s, the Hawkins Ranch House served as a special place that bonded the family. But another place of allegiance was the town of Bay City, where my mother and her siblings grew up and from which they managed their ranch by daily automobile trips into the country. My brother Frank and I grew up in the same town, as did our respective children, who were cousins, schoolmates, and friends and would become ranch partners.

In part III I describe the town as it was when I was a child in the 1930s. In both town and country I found myself being instructed. One kind of instruction is that formally announced in a building with ringing bells, but another kind comes unannounced at odd times and places. It is appropriated more by wandering around and observing than by conscious application. It is this informal kind of instruction that I mean to describe.

In the country I learned that nature is beautiful but also "red in tooth and claw." Nature's harsh side can be sentimentalized on a ranch only by omitting much of what happens there. The country was not a place for formal learning but a place to see and practice certain attitudes and arts: horsemanship; crafts; cooking; and patience in withstanding heat, cold, thirst, mosquitoes, and the tedious passing of time. The lack of commercial or theatrical entertainment in ranch life made room for the gifts of charm, inventiveness, humor, and storytelling. The "good people on the place" whom I describe were well endowed with such instructive skills.

There was a unifying moral perspective among the people I encountered in town and country. As a child, if I were corrected by other adults, I could be sure my parents would take the same view. This moral "given" was never analyzed; it came along with the set of generally accepted cultural assumptions of the time—the ethos, *as I later learned to call it. It was the same perspective encountered at the beginning of this book when in 1829 John D. Hawkins paid tribute to his grandfather. His injunctions, in paraphrase, were to renounce idleness and self-indulgence; live within your means; and make yourself more useful to others by increasing your fortune.*

This view could be labeled the Protestant work ethic, associated with Max Weber's The Protestant Ethic and the Spirit of Capitalism *(1905). The injunctions were identical to the message of Carlo Collodi's* The Adventures of Pinocchio *(1883), which my teacher Miss Tenie read to me on her porch about 1935. This perspective was assumed everywhere I went in the town and country of my childhood. The given view was unchallenged, and the widely divergent cultural options that came later had not yet broken in upon the people of the town and country. What did break in was World War II, which further unified the common values, summed up as "the war effort."*

Part III ends with a brief description of how things operate in the Hawkins Ranch partnership today, raising the question of whether the sense of place will be as central to future Hawkins Ranch partners as it has been in the past. And by place *I mean the binding power of the Hawkins Ranch acres, the Ranch House, and the town.*

Chapter 22

THE COURTHOUSE SQUARE AND DEPOT

When James B. Hawkins said he was going "to town," he meant the town of Matagorda. In his earliest days in Texas, Bay City did not exist. When the five children of Frank Hawkins were brought "to town," after the death of their mother in 1896, they came to Bay City. It was then a very new town, barely established. The Rugeley grandparents of the Hawkins children, with whom they were taken to live, were one of the first families to build a house in the new town of Bay City. In the thirties, when as a child I went "to town," I meant to the square, which consisted of the courthouse in the town's center with store fronts surrounding it on all sides. It seemed to me that everything one could ever need was located on the square. There were two barbershops, two confectionaries, two grocery stores, two pharmacies, two dime stores, two jewelry stores, two "picture shows," two banks, one cleaner, and, prominently on the corner, the United States Post Office.

Out from Aunt Janie's front porch, just beyond the sidewalk that ran along the front of Aunt Janie's house, was the town's only boulevard, Avenue G. The summer made a hot griddle of the streets and sidewalks. Barefoot, we children had to make running hops from the blistering patches of pavement to the grass. The sidewalk along the front of Aunt Janie's was mercifully shaded by a row of oak trees, planted sometime in the 1920s, the roots of which had lifted by a few inches some of the sidewalk's concrete squares. When we roller-skated along here, we had to leave the buckled part and walk awkwardly alongside in our heavy metal clamp-on skates. For my father, Jim, who walked to the bank every morning, this patch was not a hazard. No ambler when he was on his way to work that he loved doing, he swung along at a military clip. Uncle Harry had to be very careful. "I better watch my step," he would say.

Parallel to the oak trees and sidewalk, Avenue G's median was a block-long oval of grass. In the 1930s, before Victor LeTulle went out and supervised the planting of some ligustrum and crepe myrtles for the benefit of the town, the ovals were unencumbered and offered an ideal surface for children to play tag or "slinging statue." The town fathers had named the grassy median the "esplanade," pronounced "espernod" by children in my day. My own children—and yes, they played there too—called it, the "eskimo."

Bay City became the county seat of Matagorda County, one of the twenty-three original counties established in Texas in 1836 when the Constitution of the Republic of Texas was adopted. But Bay City was not the original county seat; that had been the town of Matagorda situated at the mouth of the Colorado, where some of Austin's first colonists had settled in the 1820s.

By 1894 citizens of the county, together with a land developer from Colorado named David Swickheimer, conceived the idea of establishing a more centrally located town and making it the county seat in preference to the historical but remote coastal town of Matagorda. The move required the votes of a majority of citizens of the county, achieved by the enthusiastic public support prominent citizens gave the proposal and by the promise of Swickheimer and his partners of a significant donation toward the building of a new courthouse at the Bay City location. The voters approved the move. The moving of the county seat from Matagorda to Bay City illustrates the shift that had taken place in transportation. In the 1820s when the Austin colonists arrived, creeks, rivers, and bays were the important thoroughfares. Boat transportation gave the town of Matagorda its prominence. By the turn of the century, however, railroads had become available as a mode of transportation and were heralded when they came to a town. "Blessing" was the name given to one county town when the railroad first arrived.

The Rugeley house, built in 1895, only a year after Bay City's founding, was at Third Street and Avenue H, a half block from the vacant lot where Aunt Janie's house would be built in 1913 for all the young Hawkinses.

Traffic along Avenue G was sparse because four blocks to the south of Aunt Janie's, the boulevard came to a dead end at the small railway station we called the Depot. In my early childhood Henry Cooper drove his wagon and mule team along Avenue G twice a day to take the mail to the train from the post office on the square. Every now and then frogs hopped the wrong way and got crushed under the wagon wheels so that they dried in the sun and became pieces of flattened leather as hard and clattering as tin. On our bikes we had to steer around them.

Mr. Curtis ran the train station in the daytime, and Austin Lee was the night clerk. Running the station meant sending and receiving telegraph messages in Morse code, the importance of which was emphasized to us younger children by one of the neighborhood Boy Scouts, Adelbert Verser. Adelbert instructed us on how important it was to learn to tap out SOS in case you were in danger, and Rin Tin Tin, chained up by those bad men, could not run to your rescue. Our Saturday picture shows made us believers in the Morse code.

Mr. Curtis sometimes received a message that had to be delivered to a passing train. Then he had to stand perilously close to the track holding up a long pole with a metal loop that looked like a crab net that had lost its netting. Attached to the loop was the all-important message. As the steam engine slowly chugged by without stopping, the brakeman leaned out and retrieved the message by reaching through the loop with his arm, detaching the loop and the message with it. Watching that operation made us shiver to think that if the brakeman missed the loop, or if Mr. Curtis failed to hold it up just right, there was sure to be a terrible train wreck and people would be killed. So we thought.

Sometimes we courted danger ourselves. When we heard a distant whistle blow, we were known to race out and put a penny on the track and, when the train had passed, to retrieve it, flattened and stretched out into a satisfying oval. But one day my school chum Virginia, whose father was a city constable, announced we had sure better not do that anymore because it was "defacing a coin minted by the United States government."

At the north end of the boulevard, in the direction opposite to that of the Depot, lay the town square. The courthouse was placed impressively in the middle of its block, surrounded by the small businesses of Bay City. Merchants advertised that they did business "on the square." As originally built in 1895, the courthouse was a brick building with turrets, high narrow windows, and a cupola—all the vertical elements announcing Victorian style. Redesigned in 1928, the courthouse became flatter and a little less dignified. The cupola and turrets were removed in favor of a flat roof that gave it a Mediterranean look. But at each of its three entrances it had huge carved oval stone arches worthy of an Assyrian ruler. The design of 1928 was the courthouse of my childhood, and it remained so until it was replaced by an entirely new building in 1965.

A lawn surrounded it in the 1930s, and pecan trees lined the sidewalks on all four sides. The courthouse was a symmetrical space, a pleasing sight but miserably hot in the summer for anyone doing business inside. Fans and paperweights on desktops were a necessity, and spittoons were under most desks. The courthouse was, of course, open to the public every weekday and could be entered from three sides. To cross, say, from Stinnett's confectionary on the east side of the square to Huston's Drug Store and the post office on the west, we could simply cut through the long east-west corridor of the courthouse.

One of the two confectionaries, either Stinnett's or the Alcove, was a regular stop after the picture show, where we went to see Ken Maynard or Johnny Mac Brown on Saturdays. My neighbor and schoolmate Virginia and I once happened to pass in front of the Alcove just at the time Mr. Delano was loading up a candy machine with a big blob of peppermint taffy, and we watched to see the steel arms turning. They jutted out horizontally and circled like a Ferris wheel to stretch and fold the blob. Mr. Delano, an expert candy maker, had slowed the machine to just the right speed so that as one soft looping coil began to stretch down of its own weight, a steel finger would come around to bring it up again. We watched through the store window for a long time before going inside for our ice cream cones. Walking home, with our cones softening at the bottom, we passed the *Daily Tribune* office. We waved at the elderly editor, Carey Smith Sr., who sat at a roll-top desk and wore high-top brown shoes. He always waved back.

Near each confectionary was a barbershop, the operations of which were fully on view from the sidewalk if one looked through the big glass window. Without the slightest self-consciousness, we stood and watched the barber flip out a striped sheet to cover a round stomach and take a towel from some steaming place in the wall and swirl it around a protruding nose. There was the rhythmic stropping of the razor to watch and the mounding up of the beautiful snowy lather. Once, through the plate glass, we saw a demonstration of the new baldness treatment that the Clark Brothers had just instituted in their shop. They used a piece of equipment that fitted the head snugly by means of a rubber rim and that looked for all the world like a domed papal crown made of white enamel. What was more, the device was motor-driven and rose and fell as it imparted a therapeutic suction to the scalp. We thought it the silliest thing that ever could be and quit our observations as the shoeshine man popped his rag on the boot of an enthroned pharaoh, who looked down to size up the job.

The store windows of Tett's Jewelry had an eye-catching display of what Miss Iris called silver hollow ware; and it had necklaces, diamond engagement rings, porcelain figurines of dancers, baby cups and teething rings, Add-a-Pearl necklaces, Hamilton watches, Masonic rings, and watch fobs made to look like tiny oil rig drilling bits. Sometimes we ventured inside and made a show of selecting the pattern of silver or china that pleased us most. Chantilly? Strasbourg? What about china? Franciscan!

The dry cleaner's, Verser Brothers, had a sign that read, "We dye for you." Verser's smelled of cleaning fluid but had little for us to investigate except a few signs about getting measured for a tailor-made suit of clothes the store would order for a customer. Sisk's Hardware and Grocery Store nearby had in its window an immense collection of Case pocketknives, with the three or four blades of each opened into a little metal fan. From small to very large, each knife had its own rectangular niche, and the whole collection made an impressive pattern. Inside Sisk's a huge mounted head of a steer with big horns stared out with wild glassy eyes. The store smelled of rope, of leather saddles and bridles, and Hoppe's gun oil.

The two dime stores were our special delight. There were hair rollers, pipe cleaners for artwork, watercolors, colored pencils, little pads of paper in rainbow colors, crayons, and cards of elastic. There were all sorts of candies: red hots, cream-centered chocolate drops, root beer barrels, malted milk balls, corn candy, tricolored strips of jellied coconut, and licorice whips in red and black. And there was Fleer's Dubble Bubble gum with knock-knock jokes printed inside the wax paper wrappers.

From Aunt Janie's house, we had before us a whole town with a population of about six thousand that we were free to explore. Virginia and I went all over it either on foot or on our bikes, sometimes joined by her sister Jo or by Adelbert Verser, our scientific genius friend whose father was the dry cleaner.

When summer evenings conferred the comfort and mystery of darkness, we asked to "play out" after dark. Then we left the esplanade because our games required someone who was "it" to go farther afield in search of the rest of us. We hid in dark places among prickly things and things that rustled behind stacked wood where a rat might run out—places of manageable, invited danger. If the game was Piggy Wants a Signal, one of us made a valorous dash out of hiding to wave a hand and give a prisoner renewed citizenship in the game. And after we had seen, at the local picture show named the Franklin, the 1935 version of *A Midsummer Night's Dream* with Mickey Rooney and Olivia de Havilland, we had a more magical sense of slipping around bushes when we played out on a summer's night:

> *Or in the night, imagining some fear,*
> *How easy is a bush supposed a bear?*
> (Theseus, act 5, scene 1, 21–22)

We never questioned the premise that pleasantness was already in being and that it was everyone's responsibility to preserve it. We were not unmindful of squalor, poverty, or sickness; we knew, sometimes in the most intimate way, those who were maimed or down on their luck. But we tended to regard those things as matters fated to be, matters made a little better by small acts of individual kindness. In our childhood, racial segregation was so institutionalized that we had no perspective from which to question it. When we saw separate waiting rooms at the Depot or in doctors' offices, separate schools and churches, separate water fountains; we counted the arrangement as destined to be. This was the way things already were. Fatalism and a child's trust in current authority discouraged thought of social or institutional change. All ills seemed matters to be redressed one-to-one by the kindness of individuals.

We children knew that ours were the years of the Depression because we counted the number of hobos riding on top of the freight cars passing through town. I sometimes came home to find a tired man sitting at our doorstep eating from a dinner plate my mother had given him. Virginia's father, the city constable, once let us ride along with him when he gave a hobo a lift to the edge of town to catch the next freight train out. Mr. Moore pulled his billfold from his pocket, gave the man the dollars he could spare, and wished him good luck. We watched as the man walked toward that place along the track where the train slowed almost to a halt.

Chapter 23

The Alley Way

My parents' house, which they built in the 1930s, fronted like Aunt Janie's on Avenue G and was only a block from hers—a block closer than hers to the Depot. Our house was red brick with a center hall, an early American design with the front door near the middle and the upstairs and downstairs windows positioned symmetrically. To get to Aunt Janie's taking either the boulevard way or the alley way, I had to cross only one street, Third Street, where there was almost no traffic. I made the trip often and had only to sing out, "I'm going to Aunt Janie's."

My usual thoroughfare when I was a child was the alley, which had in it all those cast-off things children valued. It was also my main route for setting out with my friend Virginia on one of our investigative missions. The alley gave us the further advantage of providing access to our house or any neighbor's house through the kitchen, a preferred entry because it plunged us immediately into interesting activities; and there was no need to ring a doorbell.

I might come home through our kitchen door just at the moment when a lump of biscuit dough needed rolling out, and I could use the rolling pin and the biscuit cutter. Sometimes a new can of Crisco had just been opened by means of the attached key to reveal its divinely smooth white untouched top; then I could spoon out the very first measure. If I came home through the kitchen door in the afternoon I might hear my mother's bridge foursome in the little library next to our kitchen. "Why, partner, those hearts are beautiful. I know we're going to make it now." Then from the oven and into me came the irresistible aroma of baking pie crust, hot caramel, and toasting meringue. The players would soon take a break, and Odessa Gatson (later Brown), who had worked after school for Lizzie and was now working at our house, might take a butterscotch pie out of the oven. Good things happened to me when I entered houses by the back door. I felt free to enter the kitchen door of any of our neighbors; no door was locked in our town except at night.

The alley was full of items we prized: a mostly empty pint of whiskey, a half burning cigarette, a profusion of unwanted but flourishing Queen Anne's lace, a roofer's kettle of hot tar, an empty Garrett's snuff box. We never touched the drops of whiskey, but we were not above taking a puff from a discarded cigarette. If we had no more important mission, Virginia and I collected Queen Anne's lace and put the stems into a jar of water mixed with blue ink. By the time we returned from the picture show, the city swimming pool, or our circuit around the square, our bouquet would magically have changed from white to blue. The empty snuff box, a shiny little round steel canister of not more than two inches, was too cunning a find to leave behind, but we had no real use for it except to hold a nickel or a marble or one jack. When our across-the-alley neighbor, Mr. Matchett, was having his garage roof fixed, we passed a barrel of melting tar with steam rising from it. The tar seemed an appealing chew: black and glistening and warm and pliable, like licorice. We pulled off a warm glob from the lip of the barrel, only to find that chewing it turned it to brittle crumbles that we had to spit out.

To get to the alley from our house, I first passed along the backyard sidewalk that ran beside our wire-fenced chicken yard. Inside were not only chickens and ducks but also a wooden hutch of guinea pigs that my brother Frank was breeding. The ducks slapped their webbed feet in the muddy puddles in our yard. In the 1930s, it was usual for houses in town to have a chicken coop and sometimes even a milk cow and shed; when chicken was on the dinner menu, I was likely to see a chicken with a wrung neck in death spasms on the ground.

Frank and his friend Louis Matchett were always growing, breeding, or hatching some form of animal life. They competed in raising game cocks and swapped eggs with other boys. Their exciting plan was to hold a neighborhood cockfight in Aunt Janie's cow lot, just up the alley. Beforehand, a terrible seriousness darkened my father's features. He did not forbid the fights, but he sternly decreed that the fighting cocks must by no means be fitted with the deadly sharp steel shanks worn like spurs and designed to kill one or both roosters. "You boys absolutely mustn't use those things. Those roosters could kill each other, but worse than that, they could fly against somebody standing on the side and cut them up badly." Somehow Louis and Frank had indeed got hold of these lethal pieces of equipment, but, admonished, they set the shanks aside. With little chance of a fight to the death, the event was to be a rooster boxing match, and the neighborhood gathered to watch.

We walked up the alley to Aunt Janie's cow lot, where the fight was to be, and stood around waiting. The roosters were brought out, each in its own small wooden box, and placed on the ground close enough together to assure that the first sight each would have would be its opponent. Their boxes, open at the bottom and placed over each rooster, were lifted off. Out they came with prideful step and stretched wings. Their feathers riffled along their necks. Extending their claws, they flew at each other with murderous intent. They flew at each other a second time, but then they did not fly again. Instead they settled into a painful indifference, and long before the spectators were ready to quit, the roosters had had enough. They folded their wings, stopped strutting, and began a quiet independent search along the ground for grains of wasted cow feed.

The alley was my route to visit our primary school teacher Miss Tenie, who also came through the alley to pay us unannounced visits; her house was in the same block as ours. Daughter of the Mrs. J. D. Holmes who had taught the five Hawkins Ranch children and Rowland, Miss Tenie lived alone, but she was gregarious, a welcome visitor who brought little pieces of news on her circuits around town. Because the alley had muddy spots, on approaching our house she scraped and stamped her feet; the shoes she wore were the serviceable black lace-up kind. In winter her entrance made us laugh because her rimless bifocal glasses fogged over in our steamy kitchen and rendered her helplessly disoriented until she took them off and wiped them dry.

"Miss Tenie!" was my father's greeting, as if her coming were the grandest unexpected pleasure. "Come have a seat with us. We've got a roast and rice and gravy and some cabbage and cornbread, and we need help with it." Taking a seat, Miss Tenie would launch into the latest bit of news. At the time of the Boy Scout Jamboree of 1937, she gave us a review.

"You know the Boy Scouts just got back from the Jamboree in Washington, DC, and I don't think a one of them got either a bath or a wink of sleep the whole time," she recounted. She said they slept in tents if they slept at all and were thrilled to see the Capitol and have President Roosevelt come and review them. "Mrs. Verser said Adelbert and the other scouts were just about worn out from the trip, but they had a fine time.

I bet there's going to be nothing but wash day for the next week." Billy Thompson had played a trick on Adelbert, slipping a four-pound flat iron into his bag. "He carried that extra weight the whole way, and never knew it!"

The boys had worked at odd jobs to help pay their way to the first National Boy Scout Jamboree, in 1937. Scouting was part of their introduction to hard work and self-reliance. Over the radio in February 1937 the president said:

> Tonight I am especially happy to renew my invitation for the Boy Scouts to hold their Jamboree in the Nation's capital. . . . Our country was developed by pioneers who camped along the trails which they blazed all the way from the Atlantic Ocean to the slopes of the Pacific. . . . Our Jamboree, besides being an event long to be remembered by the boys who participate, will afford a practical demonstration of the principle of self-reliance which Scout work is developing in all of you.[1]

Directly across the alley from our house, in the Versers' backyard, was the best climbing tree in the neighborhood: a large mulberry tree with smooth bark and spreading limbs that were easy to reach. It had big leaves that provided lovely shade for us when we sat on a branch to talk about a Tarzan movie and how wonderful it would be to live in a tree house like Tarzan and Jane and to be able to make breakfast for the whole family out of just one ostrich egg.

Adelbert wore gold-rimmed glasses that were always stretched a little too far down the flat, moist bridge of his nose. He knew all there was to know, it seemed to me, about chemical reactions, the stars, electric currents, gravity, magnets, and how machinery worked. Baking soda put into water and vinegar, he told me, would cause a fizz. "See? The sodium bicarbonate is a base and the vinegar is really citric acid," he said. "That means that they react to produce another compound plus a gas that makes the fizz."

Because an older boy genius, Morton Curtis, the son of the station master, was ready to go off to college, Adelbert had acquired Morton's ambitious collection of chemicals. It included lamp black, sulfur, saltpeter, and all the things necessary, Adelbert said, to make gunpowder. And it also contained glass vials of serious acids like hydrochloric and sulfuric, which Adelbert would not use in his experiments and demonstrations. He cautioned with the jingle:

Little Johnny Brown was thirsty
But he ain't thirsty no more
For what he thought was H20
Was H2S04

When no one of significance was available to come see some experimental effect, Adelbert would cross the alley and fetch me. Although much younger, I was nothing if not a marveling observer.

"One of the properties of magnesium is to burn with a very white light." Adelbert spoke in his instructor's voice as he took a long narrow strip out of a stoppered glass tube. It was just as he said. When he lit the strip, there leaped into life a blinding halo of white light. "Every element has certain characteristics that only that particular element has," he intoned, "and such is the nature of magnesium that when it burns, the light will be white like that."

My goodness, how enthralling it was to think that simply every common thing before us was made up of some combination of elements and that each element had certain characteristics almost like the personalities of people. "Such" is the property of magnesium. Many years later I came across this "suchness," in readings in Buddhism. It means, as best I can tell, the irreducible, trans-rational "given" of all reality and of each part of it. How mysterious it is that the world is there at all (why is there something and not nothing?) and how even more mysterious that it presents itself to us in particular ways. In any case, when Mr. Williams, in high school chemistry class, first pulled down the large shiny chart showing the Periodic Table of Elements, it was impressive to see nature—unruly and savage, beautiful and indefinable—made to present itself in rational order.

Adelbert had iron filings in a little round cardboard box, and he sprinkled these carefully onto a paper he had placed over two magnets. He pointed to the rivulet pattern made by the current as the gray filings of iron clumped about the poles. "It's like the North Pole," he said. "The charge at the North Pole is what makes a compass arrow point north so that you always know which way is north if you have a compass with you."

One afternoon Adelbert crossed the alley to our house to claim he knew how to turn a bucket of water upside down without the water rolling out. I went across to his backyard to witness the experiment. It seemed a little sneaky to me, when instead of turning the bucket upside down as if intending to pour from it, he held the bucket by its handle and slung it in a wide circle that arced from his knees up above his head and around and down again. "It's like gravity," he said. It was true he hadn't spilled any, and it was true that the bucket was upside down for a split second. But it was hardly true that he had held the bucket upside down, as in pouring. What does "upside down" mean? That's what Uncle Rowland would have asked.

By means of the alley way, there were brought to my attention little things viewed with curiosity and other things beheld, in some degree, with wonder. The latter sense eventually fused in my own psyche with the sense of awe that came into us children, coached by our Sunday school teacher, Mrs. Curtis, to be quiet, to be reverent, to be reflective when coming within a holy place, "like the Church," she said.

Much later, when I learned that Socrates held that philosophy begins in wonder, I took a significant dive into that subject and, when I taught university students, they sometimes asked, "How did you get interested in philosophy and religion and their connection?" I never had a good answer; it seemed too complicated. I needed to be able to take them through the puddles down the alley way to encounter whatever sense of wonder might break in upon them there. But would the alley way do it for them? Perhaps some commonplace of their own would be its occasion if it were beheld as an unanalyzable given.

Even then it began to seem to me that wonder simply arrives, creating an interior reverence, that its arrival cannot be coerced, and that a hard sell banishes it. Later I read enough to become convinced that this primal wonder is not the same as a "problem," which is something one has to figure out by some step-by-step process like those momentary pauses in the play of the bridge foursome when each fell silent while considering what to bid or which card to play next. Wonder is rather instantaneous, unarguable—an astonishment that overtakes one when the world or some aspect of it presents itself simply as a given. The "numinous," I later learned to call it. There in the alley way, as a child, I sensed it, but I did not yet have a word for it.

Next to the Versers along the alley came the Matchetts' house, or rather their garage. Mr. Matchett had a furniture store and undertaking parlor, a usual combination then. When Mr. Matchett left his garage doors open, we could see inside, nailed to the rafters, all his old car license plates: 1928, 1929, 1930, 1931, 1932, 1933, 1934. The numbers went back to years far beyond our memory. We counted the plates, and they matched the years of my life. Virginia said the convicts in the penitentiary in Huntsville made all those license plates and had to stamp out a new one every year for every single car in the state of Texas because it was the law. I wondered why Mr. Matchett bothered to get out the ladder and a hammer and climb up there to nail up those old plates, as if he felt that a year, when it was all finished, was still too valuable just to throw into the garbage can. Maybe he cherished the year by nailing up the license plate, cherished all the years he nailed up.

But there came to the Matchetts the saddest of all years. Their son Louis contracted tuberculosis. Members of the bridge foursome said they bet it was because Louis had taken a summer job in the ice house, and going in and out of the cold storage locker was bound to have weakened him, tall and thin as he was anyhow. There was no effective treatment at the time beyond bed rest and scrupulous attention to hygiene to prevent infecting other members of the family. The foursome spoke sadly and admiringly of how courageous the Matchetts were. "Mrs. Matchett washes his clothes separately and keeps all his silverware and dishes separate. And she has to wash down the floor and walls with a solution of Lysol."

Weeks went by with Louis in isolation, lying in his bed by the window, across the alley from our house. One morning as we sat at our breakfast table, we saw Mrs. Matchett coming up the backyard sidewalk, a slender lady with sandy hair parted in the middle and drawn to each side in pinned-up braids. She came in and stood just inside the door, and my mother rose from the table to meet her at the door. Mrs. Matchett spoke quietly to my mother, and when she left, my father shook his head. She came not to say that Louis had died but to ask if my brother, who of necessity had not visited Louis in many weeks, could possibly now come over to see him. It would be a last goodbye.

Chapter 24

Miss Tenie

My almost daily trips up the alley for visits to Miss Tenie made me her pet even before I started to school at Jefferson Davis Grammar School. She had a screened porch at the front of her simple one-story wood frame house, and on it was a two-seater porch swing, the kind that in movies provided the place of budding courtship. In her late fifties now, Miss Tenie Holmes had never been married but was anything but spinsterish. My mother said she had once been in love with a gambler, but "that wouldn't do." She also said, "You know, when Miss Tenie was young, she got put out of the Baptist Church one time for dancing!"

She was like a grown-up playmate and seemed genuinely to enjoy and be stimulated by the presence of children. She was my first grade teacher and taught hundreds of others who preceded and followed me. Sometimes she let me help her make copies on her hectograph. Using special ink, she traced over a line drawing of a circus elephant and then laid the tracing sheet flat and smooth on a shallow tray of light gray gel. When she lifted the sheet, there the elephant was, outlined in blue on the gel itself. Then she showed me how to smooth a blank sheet of paper onto the gel and print the elephant on the paper. She saw that I wanted to stick my fingernail into the gel, the way I did in biscuit dough, but told me that was not a good idea.

Often I sat in her porch swing beside her while she read to me. There was *Pinocchio,* the story of the wooden puppet constructed by his "father," the carpenter Geppetto. Pinocchio yearned to be a real little boy instead of a wooden counterfeit one, but to become a real boy required him to have good character: to tell the truth, be respectful to his father, go to school or to work, and be generous and kind instead of lazy and selfish. Pinocchio failed miserably in all these expectations. He rudely snatched the wig from his father's head and refused to go to school or to work. And with every lie he told, his nose grew longer. The wise cricket advised Pinocchio to learn a trade, but Pinocchio wanted to be a vagabond and amuse himself all day long. A selfless thought finally entered his head only when his beloved, beautiful guardian fairy fell into desperate need, and he wanted to help her. But without any means, he wondered how he could help. Suddenly a single thought transformed him: "I can work!" Miss Tenie did a school teacher's pounce on the moral of the story: Pinocchio left his lazy ways, became helpful, and all of a sudden changed into the real boy he wanted to be!

From Miss Tenie's porch I waved at the Kogutt children, who lived directly across the street. Sarah had taught me to roller skate right there in front of Miss Tenie's, and Samuel and Albert Kogutt were in the Boy Scouts with Adelbert. For some occasion, Miss Tenie had once gone with Samuel to the synagogue, and after that she laughed at herself for having pestered him to take off his hat. She nudged and nudged him and whispered that he ought to remove his hat in a place of worship. Well, she told me, in the Baptist church the men took their hats off, but in the "Jewish church" they were supposed to keep them on. All I knew about differences in worship was that Methodists and Baptists chatted a lot when they went into church, but that we Episcopalian children were directed to be silent, go to our knees, and say a very sincere prayer on entering God's house. From noticing how Miss Shirley did it, I assumed that the sincerest attitude required closed eyes and the use of thumb and forefinger to keep a soft little pinch in place at the narrow bridge of the nose.

Miss Tenie's real first name, Kathleen, was known to almost no one except my father. She had been a "teeny" baby, she told me, and so was always called Tenie, somehow losing the extra *e*. She was born in 1874 and as a girl had briefly assisted her mother, who taught the Hawkins children in my mother's generation. But her most vivid memory of making a start as a teacher had to do with accepting an assignment to teach the children of the keeper of the Saluria Light. This lighthouse was in a remote spot on Matagorda Island, south of Port Lavaca, across from Pass Cavallo. The year of her first teaching assignment was 1890, her sixteenth year, and the children were those of the Hawes family, related to the founder of Saluria, Judge H. W. Hawes.

In the 1850s Saluria was a busy Texas port that served ocean-going ships as well as local traffic across the Matagorda and Lavaca bays to the towns of Indianola and Matagorda and beyond. In the port's heyday, mail boats called regularly and a stage coach provided service to and from ships. J. B. Hawkins had noted in his pocket memorandum book shipments of sugar to Judge H. W. Hawes at Saluria. There were warehouses and a wharf to accommodate busy commerce in Saluria before the Civil War. In 1862, in order to keep the port from falling into Union hands, these warehouses and other installations were destroyed, and after the Civil War there was no appreciable rebuilding. A few stalwart souls hung on and made a modest rebuilding effort; prominent among them were members of the Hawes family. As if the post–Civil War years were not dismal enough, there next occurred two devastating hurricanes, one in 1875 and another in 1886, which finished off Saluria and nearby Indianola as commercially viable ports. A letter written by E. Hawes Sr. following one of the storms gives the sad details.[1]

Thus Miss Tenie made her professional start at a place that, for the most part, had come to its end. It is hard to imagine a more discouraging beginning, but in fact her first day was still more discouraging. She enjoyed recounting the story. A relative in the Hawes family had died just as the new teacher arrived, and Miss Tenie, the only person on the island not related to the deceased, was called on to bathe and prepare the body for burial. If that was what the job required, that was what Miss Tenie did.

Despite that first grim experience, Miss Tenie's desire to be trained as a teacher was not in the least dampened. She attended Baylor University for enough sessions to become accredited as a teacher and taught with characteristic zest in elementary school until her death in 1952. A school built in Bay City that year was named for her, the Tenie Holmes Elementary School.

My trips up the alley to Miss Tenie's changed somewhat after my father decided, around 1936, to build a better garage than our old wooden one. His decision had the unforeseen consequence of encouraging another visitor to spend time with Miss Tenie.

My father planned a brick garage with hinged wood doors that opened vertically, spring assisted, though not motorized. He wanted to replace the old garage, which had double doors that opened on hinges fastened to each side of the door frame. These doors sagged on their hinges, making it a struggle to pull them open, as they dug into the shell gravel. Most people just gave up and left their garage doors open. My father thought the structure of the old garage was sound and that it might make a small apartment for a tenant—a bachelor, for instance. He thought it might be fixed up for that purpose, and he proposed having it moved up the alley as a gift to Miss Tenie, so that she could have a tenant and another small source of income to supplement her teaching salary. As a town banker, my father was given to thinking up possibilities of this kind.

When the men who moved houses by jacking them up onto a platform came for our old garage and moved it up the street to Miss Tenie's backyard, she had the good local carpenter, C. K. Norcross, fix it up with a bathroom and a small kitchen. "It turned out real nice," my father said. "Very suitable and economical for a young bachelor." When the insurance man Ludolph Heiligbrodt moved to town, he stayed in that apartment until he and his sweetheart Gladys decided to get married. Then Norcross himself rented the space.

After that, when I visited Miss Tenie in the afternoons, she and Mr. Norcross were sometimes sitting on her porch having a small wedge of rat trap cheese and drinking iced tea from huge, footed, round-domed glasses. Once they both had beer in their big round glasses and gave me a taste. First, though, Mr. Norcross said that if I wanted to taste it, I might like it better if a little salt were sprinkled into it "to cut the bitter taste." I wondered why salt would make anything taste less bitter but felt sure I could find out if I asked Adelbert.

Even though I knew Miss Tenie well, the morning the other new first graders and I were directed to sit down on the concrete front steps of Jefferson Davis Grammar School and wait a few minutes for the bell, my stomach clenched. When we filed into the hall that smelled of newly applied floor oil and then into Miss Tenie's classroom, she began by pointing a long stick at the letters above the blackboard. In the days to come we would sound them out and get started on *Dick and Jane.* At noon, after we had burrowed into our lunch sacks, Miss Tenie spread big squares of newspaper on the floor and we each took an oil-cloth-covered pillow and lay down for a nap. "Curtains down," Miss Tenie said, to tell us to close our eyes. Only Rady, the smallest boy in class, went to sleep, and he stayed asleep even after Miss Tenie said, "Curtains up!" We laughed and wanted to wake him. "Oh, let him sleep," said Miss Tenie.

In the 1930s when we started to school, most families did not move often. That kind of scattering in pursuit of work opportunities grew common only after World War II. Before that, chances were that you would remain through all the grades with many of the same children who had begun first grade with you. In Miss Tenie's room I happened to be assigned to the same table as Roland Bussell, a sturdy and reliable little boy. One day when we saw through the window a dark sky and rain coming down in silver streaks, recess outdoors was not an option. Miss Tenie got out the Victrola and put on a spirited Sousa march. Two by two we marched around the room, stamping our feet. The Victrola was not one that plugged into the wall; it had to be cranked by hand and had no mechanism that allowed the one who turned the crank to take any time to rest. If the crank stopped turning, the music also stopped. Miss Tenie soon began to tire of the repetitive motion and asked Roland if he could help her by taking a turn.

Oh, could he! Roland's face lit up with incandescent joy at this responsibility. You bet he could.

More than sixty years later, at our fiftieth high school reunion, I reminisced with Roland, now a retired engineer. "You have never lived," he said, "until you have been asked to crank the Victrola in the first grade!"

The stability of the schools, the teaching profession, and the student population meant that we all grew up together and had common experiences and common memories year by year. There was little expectation of dramatic change. We were thus unprepared for a sudden new development in Miss Tenie's life. One day I was in the kitchen and overheard the bridge foursome in a state of consternation during a break from their game.

"What on earth could have possessed Miss Tenie? Why in the world would she . . ."

"Well, you know Miss Tenie. She always had a venturesome spirit—you know she was engaged to a gambler once. I think she just wants to try everything."

"Well, there's no way to tell how this will turn out. I sure wouldn't bet on it. She's going to have plenty of unexpected turns now. She just doesn't have any idea what she's getting herself into. I sure wouldn't want to do it. Not at her age." I went into the library, and Lurline Wadsworth turned to give me the breaking news. "Margaret, did you know Miss Tenie is going to get married to Mr. Norcross!"

My mother had been the first to know. They were to be married before Judge Styles at the courthouse without any ceremony. My mother pressed Miss Tenie into having the marriage take place in our living room so that she could serve refreshments and have a small gathering of close friends. The date of the marriage was to be Wednesday evening, December 30,

1936. My mother asked Lizzie if she could strum the wedding march on the Mason & Hamlin in our living room; relations with Lizzie were cool, but the sisters were not yet estranged.

When the day came, those in attendance were Judge and Mrs. Styles, Aunt Janie, Daughty and Rowland and their daughter Martha, Aunt Sister and Uncle Esker, and Rowland's sister Aunt Dolly and her daughter Betty Jane Doubek. My cousins Martha and Betty Jane and I sat on a huge ottoman in our living room as we waited for the ceremony to begin.

Mr. Norcross, wearing a suit and tie and a red face, stumbled into the house through the south porch door, a door that on a December night like this was meant to be closed for the season. A whiff of bourbon spirits drifted in. My father and Uncle Esker greeted the groom with some seriousness and escorted him outside for some walk-it-off open air therapy. The therapy worked, and the marriage took place, performed by the Baptist pastor, the Reverend Odis Rainer.

Miss Tenie had to stop teaching. A single woman was self-supporting, but a married woman had a husband as the family breadwinner; the regulations of that time required the school board to deny teaching jobs to married women. Victor LeTulle came to Miss Tenie's rescue by giving the school district the little one-story house that the Matchett family had now left, just a few steps down Avenue H from Miss Tenie's home. There, under some arrangement LeTulle made, Miss Tenie held kindergarten classes. Back in his bib overalls, Norcross busied himself with the construction of playground equipment. He built a set of swings and a big fortress-like platform with a small slide and a large one. Two wooden seesaws and a sand pile completed the grouping. Inside the house he installed blackboards in what had been the living room and a bedroom. Miss Tenie acquired an upright piano.

Norcross traded in his old Model T and bought a new tan Chevrolet coupe. To the wonderment of the bridge foursome, Miss Tenie began to take driving lessons from Buck Blaylock, a master mechanic at Rugeley Motor Company. She had never driven a car. In all her years of teaching she had walked every day to and from Jefferson Davis Grammar School, a distance of about two blocks. She had a good raincoat, galoshes, and a large black umbrella, which she shared with me when needed because we took the same route home. Daughty had one of the first sightings of Miss Tenie at the wheel: she passed by slowly, went around the block, and passed again, waving. Daughty waved back. Miss Tenie came by once more, waving ever more earnestly—she had forgotten what Buck had said about using the brake and the clutch to stop the car.

Miss Tenie was sixty-two when she married Mr. Norcross. His crotchety ways returned, as did the pint bottle. They divorced, and he left to find work elsewhere. A year or two later my father reported encountering him briefly on the square, where Norcross explained: "There were just too many old maid school teachers around here."

Miss Tenie returned to her first love, children, and continued to teach her kindergarten and first grade classes in her little school on Avenue H until she died in 1952. Her school was then taken over and operated by Virginia LeTulle Peden, who named it the Peden School. My children and my brother's attended as kindergartners.

Chapter 25

Good People on the Place

When the Hawkins sisters undertook management of the ranch, Janie assumed the duty of attending to the cattle. Meta was in charge of what was called "the farm," a strip of corn, cotton, and sorghum fields along the west bank of the Liveoak Creek, the easternmost boundary of Hawkins Ranch land. Where Lizzie failed in her attempt to manage land and cattle on her own, her sisters would succeed, although they might very well have failed too. They attributed their success largely to having "good people on the place." Janie and Meta kept the ranch productive by paying close attention, by not spending much during the Depression years or falling into debt, and by relying on the help of people who knew how to do the hands-on work.

From 1915 through the 1930s Dode Green was the foreman supervising the work of the cowhands. He could hire them from among the tenant farmers for seasonal cow work and the special bigger jobs of branding in the summer and rounding up and shipping out calves and old cows in the early fall. Dode and his wife Susie and their sons lived in the Ranch House yard in a small white house beside the bois d'arc tree. He was a short man of about five feet, six inches, with legs bowed from years of riding and working cattle. He habitually wore a red or blue bandana around his neck. Dode was clearly in charge when he and the cow hands were on their cow ponies herding cattle or when he stood on the platform at the pens helping to prod animals through the chute. He was gentle but authoritative and businesslike in giving direction to his hands.

When he rode out alone to inspect fences, he took a rifle in a leather scabbard fitted to the left side of his saddle; the barrel of the gun slanted toward the ground beneath the stirrup leathers. Another small leather scabbard held a fence-mending tool that could pull staples, hammer in new ones, and cut wire; it was an "all-in-one." And he had a coiled rope ready, hanging on his saddle by a leather thong, in case he needed to catch a calf and bring it in or throw it down on the prairie to doctor a wound.

When Dode dismounted his cow pony, he went straight into his house to remove his boots and put on comfortable house shoes. Like all cowmen, he had a ready bootjack to hold in place the heel of his boot while he slipped his foot out. Off his horse and out of his boots, Dode's bowed legs were prominent, and after work he chose to wear his house shoes even as he walked around the ranch yard. He rarely appeared without his hat. Dode smoked a small pipe with a short, curved stem, filling its bowl from a cake of Brown Mule chewing tobacco that smelled of molasses; with his pocket knife he cut a small plug just the right size for the bowl.

He had a merry and delightful way with children, who looked forward to visits with him. "Now, Missus, you better not go too close to that horse's hindquarters lessen he take a notion you a varmint and he decide he need to kick. You best go way round him." He saddled a horse named Baldy for me with my child-sized saddle, and after boosting me up, he adjusted my stirrups to the right length, a job that meant lacing a leather thong through many eyelets, not just rebuckling a buckle.

As a child I followed him around the yard. Once he knelt down beside the water hydrant at the corner of the Ranch House, which he called "the big house," to show me what a cowboy could do if he were out in the pasture, lacked a cup, and needed to drink from a windmill spout. He took off his hat and used the curved up brim to catch the water and then sipped from its front edge.

His wife Susie was also short and a little stout. She took care of chickens, turkeys, and guineas and cooked for the family and the ranch hands when they were working cattle. She rolled out her biscuits with a wine bottle. Dode's salary in 1915 was seventy-five dollars per month. Susie was paid fifty dollars per month for cooking for the hands. Living with them in the Ranch House yard were Tibby, Preston, and Froggie Wyche, who were my older playmates when I went to the ranch and who worked as cowhands as needed. One day Froggie nailed together pieces of two by four to make stilts for us. Tibby plaited strips of corn shucks to make a new seat for a failing wooden chair. In the late 1930s and 1940s, as Dode began to age, his older son Arthur became his assistant, eventually taking over the job of foreman.

The sharecropping system of farming that developed in the aftermath of the Emancipation meant that small, individual family farms, scattered and separated from one another, were worked by tenants, who in that way could support themselves. The tenants also took part-time or seasonal work as ranch hands. The ranch work needed extra hands mainly in two periods, from about April through June and again from September through October. In the spring the early calves were earmarked and branded, and the gathered herd was run through a narrow vat filled with creosoted insecticide; later a pressure spray and liquid insecticide supplanted the dipping vat. Fall was a time for more marking and branding of summer calves and the sale of marketable calves and "culls."

When the young Hawkins sisters took over in 1917, they drew part-time ranch hands from those who also farmed on shares. For cow work such hands were paid a wage of about one dollar per day, and they were furnished a horse to ride and three square meals a day during the work period. Even as the 1930s advanced, the routine work on the ranch had changed little from the time of Frank Hawkins in the 1870s and 1880s.[1] As in his day, if disease hit the herd, it was usually addressed only if it became an epidemic. Occasional deaths within the herd were to be expected, and buzzards would be seen circling above a carcass on the ground.

Sometimes loss struck harder. One winter day in the late 1930s, the wind turned sharp and the rain that followed turned to sleet. Sister put on her fur coat and Janie, Harry, Meta, Jim, and Esker donned their warmest coats and their boots, and they all went to the ranch to take sweaters and extra slickers to the ranch hands. Dode and the cowboys stuffed the tops of their boots with newspaper. Each put on two shirts and a sweater and covered himself with a long split-tailed slicker that straddled his horse. There were then no wind breaks or feeding stations to protect the cattle on the open prairie. As best they could, Dode and his hands pushed the cattle into the shelter of the woods along Liveoak Creek. They put out hay and salt lick blocks and broke up the ice that had formed on the surface of the water troughs. The weather was so severe that some cattle were lost despite the precautions.

Dode and Susie continued to live in semi-retirement in the Ranch House yard even after their son Arthur took charge as foreman. Arthur and his wife, Emma Robbins Green, made their headquarters at the Sheppard Mott. Their small T-shaped wooden house was one still in service at the location that had once been the homestead of Abram Sheppard, from whom J. B. Hawkins had bought this and various other pieces of land.

The presence there of a few scrubby trees gave this upper pasture of some three thousand acres its enduring name, the Sheppard Mott or simply the Mott.

Although Dode was retired, he and Arthur could still spend time together and did some hunting. To give themselves a little excitement and reduce the hazard on the prairie, one of their undertakings was to hunt down rattlesnakes. Arthur told a *San Antonio Express-News* reporter, "Rattlesnakes are funny things. They'll hole up with skunks. We would hunt for a skunk and when we would go to dig it out, we would find rattlesnakes bunched in the same hole."[2] Arthur and Dode said one day they had killed eighteen rattlesnakes. Aunt Janie and I, arriving at the ranch shortly after the hunt, saw at least ten of them hanging side by side on the whitewashed fence at the Ranch House.

Arthur's duties were much like his father's. His time of service coincided with my high school years, the period just before and during World War II, before my brother Frank took over the management of the ranch about 1948. At the Mott a new shipping pen with a loading ramp had been built. Arthur and his men now gathered the cows and calves there in a fenced triangular trap that led them into the pen and ultimately along the narrow cutting chute to separate the calves from their mothers before loading and shipping. The main difference from Dode's day was that instead of the animals going into box cars at the rail yard in Wadsworth, cattle trucks could now be brought to the gathered cattle on the ranch itself.

As a school girl I was often at the Mott with Aunt Janie, riding beside my aunt and Arthur as we pushed the herd along. Sometimes we sat in Aunt Janie's car with the doors open as we waited for the herd to get sorted and the cattle trucks to arrive. After the calves were separated from their mothers, they spent the rest of the day bawling out their mournful objection to the separation.

The ranch hands all knew one another well, and many were kin. In their work together they were full of horselaughs and hurrahing. When a calf dashed from the herd and had to be chased, a hand would holler something like: "Look out! We got us a rodeo now. You better mind out! That little dun one is gone to Jericho if you don't kick up that mare no quicker than that!"

The showiest and most athletic rider was Harry Gatson. If a calf bolted, Harry shot out at breakneck speed to turn the calf back. Then he pulled up his horse in several jolts before the final stop, his elbows flapping mightily with each jolt. That stylish run and elbow-flapping stop was Harry's trademark. Even from a great distance everyone knew when it was Harry who went after a calf.

The best cutting and calf-chasing horse on the ranch was a handsome sorrel named Socks for his four white feet. Arthur made a pet of Socks and had taught him to "stretch" when a rider, ready to mount, tapped his boot against Socks's fetlock. Socks would step forward slightly with his front legs only, to make a lower, easier mount for a rider.

A Houston Fat Stock Show and Rodeo man came to the ranch one day looking to buy some colorful spotted cows as rodeo animals—the wilder in temperament, the better. The Hawkins sisters laughed that the Hawkins Ranch was well equipped to accommodate that need. Their herd consisted of range cattle of mixed breeds, colors, and temperaments; these were not pure-bred cattle of uniform color, like the Santa Gertrudis that came to be a favorite on other ranches. Arthur's account of the conversation with the Houston buyer was given in 1986 to a reporter for the *San Antonio Express-News:*

> Janie Hawkins was the boss lady. She was real nice. One time we were loading up some cattle to send to the rodeo in Houston and she was up on the chute.
>
> I had this little sorrel horse [Socks] shod all the way around so he wouldn't slip, and he could really lay 'em down. Two heifers broke out and we took after them. All you could see was dust flying up in the air when that sorrel horse turned 'em.
>
> That man from Houston said to Mrs. Hawkins, "Let me take that horse to the Houston rodeo, I'll get you $500."
>
> Mrs. Hawkins told him, "I don't need the money, and if I got rid of that horse, I would have to get rid of Arthur, too."

Arthur told the reporter that the work on the Hawkins Ranch "had to go on, rain or shine." When he got a Sunday off, he said, he would go to the rodeo in Bay City: "I tried a little bit of bull riding, bull-dogging steers, roping calves. I never was much good at it. I would foul the calf. The loop would be too big, and the calf would stick its foreleg in there."[3]

After the cattle sale Janie decided to attend the Houston Rodeo with some friends and watch the colorful Hawkins Ranch cattle in action. As she sat in the vast audience, the rodeo announcer introduced her over the loudspeaker and a roving spotlight found her in the crowd. She took a bow.

The farm lands that Meta managed—the fenced acres for row crops near Liveoak Creek—were farmed before and during World War II by Rose and Douglas Chapman. They lived there, within an easy walk of Rose's parents' house. I rode along with my parents when they went to check on the corn, cotton, or sorghum. Nothing could have been more discouraging.

The corn crops were routinely a half failure, not from lack of effort but because of the poor drainage in the fields. Every season's corn had blighted rows, and sorghum showed stunted patches where water stood. Farming was discouraging unless weather conditions were almost perfect; and cotton was never an ideal crop for the coastal climate. Moreover, cotton had to be poisoned with arsenic against the boll weevil. For this dangerous duty, Rose and Douglas and other farm hands wore big straw hats, gloves, and bandanas that covered mouth and nose. They went down the cotton rows with buckets of arsenic powder, each dipping a rag on the end of a stick into the powder and then shaking it over the cotton to send poisonous particles into the air to settle over each plant.

Sometimes we extended our auto trips to the most isolated place on the whole of the Hawkins Ranch: Harkell Bend.[4] In our family Harkell Bend stood for a place of legendary remoteness and mystery, a place where one could hide from the world and never be found. During World War II Uncle Harry had a saying that if Hitler ever came over here, "We can all just go to Harkell." In Harkell Bend James Parks and his family farmed cotton and corn on shares. James was a short, muscular man with a ready smile. He had many children, providing hands to pull corn, pick cotton, milk the cow, feed the chickens, and slop the hog. The family could subsist. They could fish and shoot squirrels and possums, and on a Saturday trip into town they bought grits, lard, flour, cornmeal, and molasses.

The rampant mosquitoes were a perpetual trial, addressed by keeping a "smoke" going—a small fire was kept smoldering just outside the doorway. In the intense heat of summer, the woods gave shade but also raised a barrier to the breeze and held in the moist heat. James Parks's small children cooled off by taking a dip in the cattle trough. Once, driving into the woods with my parents on a hot Sunday afternoon, we reached James's house at Harkell Bend and happened upon a refreshing vision of smiling little round faces bobbing above the trough's water line.

Sixty years later I happened to stop at a Bay City gas station, filled up, and had a chat with the proprietor, a distinguished-looking African American gentleman, short and muscular. He was dressed in business attire because he sold not just gas but also insurance and annuities, through American General Insurance Company in Houston. "You and I, we go back a long way," he said, mistakenly calling me by my mother's name. He was Herman Parks, one of the little bobbing heads.

James Parks's house in Harkell Bend was a typical sharecropper's house. In such a house the only heat in winter was a wood fire either in a fireplace or in a cast-iron stove. There was no electricity, and the bathroom was an outdoor privy. The plank walls of tenant houses had no insulation, and when a stiff norther blew, cold wind searched out the cracks and chilled the air away from the fire. To ward off drafts, a common practice was to paste newspaper on the interior walls. They became an expanse of newsprint, headlines, and photographs, none of which provided much help in insulating occupants against the winter winds.

When a tenant like James Parks had harvested his crop and it was sold, it was Meta's job to calculate the dollar amount to be allotted between tenant and landlord. The tenant who had worked his field and harvested his crop often hoped for more dollars as his share than the current market price might allow. Meta worked at her hutched wooden secretary every fall, calculating the crop receipts. Often she had to take special pains to review the numbers with a tenant to assure him that the accounting was correct.

Arthur remembered Meta investing five hundred dollars for two hogs to raise some shoats. The boar developed an abscess on his foot, and the veterinarian came out to the farm. Being the overseer of livestock, Arthur got the unpleasant task of medicating the hog's foot: "I would take syringes, alcohol, and cotton and do all that vaccinating. . . . Cattle, horses, sheep, mules, turkeys, chickens—I took care of them all."[5]

Rose Franklin Chapman was a slender, energetic woman, an entertaining storyteller, and a much-appreciated mimic. Her father, Amos Franklin, and her brothers, William and Jim Franklin, farmed and worked as cowhands. Rose and Douglas farmed for themselves and oversaw others who farmed on shares.

One year Meta put Rose in charge of dispensing an allotment of seed corn to each tenant farmer and had cautioned her that it was poisoned to discourage weevils. Rose loved to play her authoritative school mistress role as she warned the much younger James Parks like a wayward schoolboy. She acted out her narrative, quoting herself:

> I said, James. Now you listen to me, James. Don't you be feeding your horse, mule, chickens or none of your children this seed corn. It's got poison in it! You hear me, James?
>
> James, he just shaking his head yes, and goes off carrying his sacks of seed corn, paying me no mind. Pretty soon, in a week or two, here comes James, riding up on his jolting ole mule. And I say, Uh huh! I see what happened, James. You fed that horse some of that seed corn and you killed him, that's what! I told you about that seed corn!

Just before World War II, when Arthur had taken over from Dode but lived at the Sheppard Mott, Dode's house in the yard of the Hawkins Ranch House became vacant because once he retired fully, Dode wanted to move from the country. The Hawkins sisters then decided to phase out farming and have Rose and Douglas move into the house where Dode and Susie had lived. By now the Ranch House had been renovated, and the sisters wanted someone living nearby. Rose and Douglas were glad to be finished with farming. "Over at the farm, we had our bitters and our sweets," Rose said.

In the summer we went crabbing. Members of the extended family, young and old, walked out on the wharf on Lake Austin. Children were delighted to slip a net under a big blue crab as it gnawed on a meat scrap tied to a string. A good catch would fill a big wooden box with an angry gang of clicking, pincer-menacing crabs that we boiled for our supper. Back in the Ranch House yard, with the help of Rose and Douglas, we dumped the crabs into a tub of boiling water; then, sitting at the wooden tables in the barbecue house, we each picked out our own crab meat, using a Coke bottle to bash in the shell.

The children who came into the family in the 1950s—my brother's and my own—were entranced by Rose's acted-out narratives. She told stories with a wagging finger, a swaying of her slender waist—whatever the tale required. Rose had a special performance of an onomatopoeic fable in which all the barnyard fowls were hiding in the woods in fear for their lives. She told of a big country church convocation with lots of guest preachers, and everybody knew "all the preachers do is eat up all the fried chicken and turkey dinners the church ladies can put out." As all the chickens and guineas and turkeys hid—"not wantin' to be put on the platter"—they began to call to one another, wondering if it was at last safe to come out. Here Rose stood with her hands cupped to her mouth to call out like each frightened fowl:

> *The Rooster calls,* Is the Associ-a-a-a shun broke?
> *Then the guinea way up on the roof calls,* Is the preachers all g-o-n-e?
> *And the hen says,* Not chet. Not chet.
> *And the turkey gobbler says,* Couple of 'em here yet. Couple of 'em here yet."

Rose's father, Amos Franklin, was in Dode's time a senior member of the cowhand crew. He wore his felt hat with a pyramid crimp at the crown like General Pershing's and gave a deep-voiced command to the cow dog that got in the way at the cattle pens: "*Way* from here, Dog!" Rose's brothers, William and Jim Franklin, farmed and worked with their father in branding and marking calves.

On cold evenings Rose and Douglas liked to sit beside a slow-burning log in their fireplace at the little white house in the Ranch House yard and shell pecans from the native trees that grew in the woods. Sometimes at weekends Rose would saddle the gentle horse Baldy to go and visit her brothers and their children. Douglas was older than Rose and began to be troubled with arthritis and other old age maladies. After he died, she left to live with her daughter in California. There she was briefly discovered by the television world when her daughter persuaded her to audition for a Kool-Aid commercial. The casting director was looking for "an old timey black woman." "That's you, Mama," her daughter said. And for a time Rose could be seen on national television touting the refreshing qualities of Kool-Aid.

Even in recent years, some of the people working in the Hawkins Ranch orbit did so for life. Harry Gatson's sister Odessa was a faithful servant, a gifted cook, and a deeply loved friend. Her presence in our family was continuous from the 1930s until her death on February 16, 2002. Odessa was physically large and strong, like her brothers Harry and Enoch, who were both expert cowmen. As a girl she worked for Lizzie, then later for Sister, and finally for my mother. Her incorporation into Hawkins family life was initially at recreational times—Thanksgiving, Christmas, and suppers after hunting or crabbing.

In my parents' household Odessa prepared meals, catered the refreshments for the bridge foursome, and rocked the new babies who began to arrive in the 1950s and 1960s. A jingle she recited for the children in her care, "Mary Mack," must have been a jump rope song, its rhythm calling up the motion of children on each end of a rope they turn for others to jump:

Mary Mack dressed in Black
Twenty seven buttons up and down her back
Singin' lu la, lu la, lu la
Singin' lu la, lu la lay
Singin' lu la, lu la, lu la
Singin' lu la, lu la lay

Old Uncle Ned fell out the bed
Struck his head on piece of cornbread
Cornbread rough, cornbread tough
Good God A'mighty had cornbread enough
Singin' lu la, lu la, lu la
Singin' lu la, lu la lay
Singin' lu la, lu la, lu la
Singin' lu la, lu la, lay

When the babies became toddlers and the family went to Matagorda Beach, she stood barefooted in the edge of the surf to watch over them, lifting her skirt to escape the waves. On the crab wharf, she delighted in a good catch. In the 1970s, as my mother grew frail, Odessa gave her every care in helping her learn to walk again after a fractured hip. And when other maladies could not be turned away and my mother's end came on January 7, 1975, Odessa's presence was a great comfort to my father, my brother, and me.

She was one of three children born to the farmer Moses Gatson and his wife Narcissus Edison Gatson. When Odessa, Harry, and Enoch were growing up, they lived on Buckner's Prairie just north of Hawkinsville, in a two-story house belonging to Jim Bruce. One day Harry decided to take a look over the protective banister on the second story porch. He crawled up and over it and fell to the ground, striking his head a hard blow. Harry was treated by a woman doctor from Brazoria County, but the accident had done permanent damage, robbing him of articulate speech and hearing, although not of energy, physical strength, or intelligence.

All who worked around Harry admired him for his good nature and his athletic ability as a cowboy. Even though he was unable to form words, he had ways to communicate. For instance, when the ranch hands learned in 1949 that my brother Frank was to be married, Harry greeted him with grins; he pointed a finger at Frank's chest—"You!"—and then he took a few steps with his elbow out, miming a groom escorting his bride.

As children, Enoch, Harry, and Odessa worked at picking cotton, pulling behind them along each row the long cloth cotton sacks that grew heavier and heavier as the pickers moved down the row. Odessa said Harry was the best picker, his day's pickings always weighing the most. His go-ahead spirit and skill as a horseman made him a much appreciated member of the cowboy crew on the Hawkins Ranch, and he spent his life there. He lived in one of the small white cabins placed in the yard when J. B. Hawkins built the Ranch House; the cabin is still called Harry's House.

World War II made an ugly attempt to disturb Harry's world: Aunt Janie received a government notice that Harry Gatson needed to report to the draft board. He would not be eligible to serve, but apparently he needed to report. Aunt Janie claimed to be conversant with Harry's sign language, but interpreting this government summons was a challenge indeed. Her ineptitude at it was the source of some amusement to Meta's bridge foursome. Meta told them Janie had shown Harry the letter and elaborately enacted a soldier shooting, followed by gestures of signing a document. Harry wagged his head. The next day he hitched a team of mules to a wagon, went over to the bottom land, and brought back a load of firewood. "Well," Meta said, "We could always use the firewood."

Odessa and Narcissus developed an extensive set of symbols, almost a private sign language, for communicating with Harry. Arthur needed this too, working with Harry every day taking care of the cattle. Arthur could ask Harry to ride into the pasture and bring in a particular cow and calf. He managed the message by pointing to a chip of old wood to indicate the color of a brindle cow; if the cow was a Hereford, he pointed to a shard of red brick and then made a circular motion in front of his face with the palm of his hand, to indicate a white-faced Hereford. A hand brought low to the ground after the cow gesture meant Harry was to look for a mother cow with a small calf.

Harry was not required to serve in the army. In later years the Hawkins

Ranch books revealed that Harry had been given numerous salary checks and had cashed hardly any of them. His wants were few in the little cabin in the Ranch House yard. Replacing all his uncashed checks, the Hawkins Ranch opened a savings account for him at the bank.

One day in 1970 my husband provided Harry a new adventure. My brother Frank had leveled part of a pasture in front of the Ranch House to make a hay meadow, and next to it he created a landing strip for a light airplane. My husband had a small plane at the time and enjoyed landing there in good weather. On one such visit Harry walked up to the plane and pointed to the sky with a twirling hand, obviously asking to be taken up for a plane ride. For the first time he saw spread out beneath him the land over which he had so often ridden his horse and turned calves back to the dusty herd. Surprisingly calm, Harry viewed the sight as if it were the most natural thing in the world.

A good all-round hand on the Hawkins Ranch was Cleveland Woodward. He was not one of the cowboys; he liked farming better than working cattle. Cleve's first duties at the Hawkins Ranch were to help out with cutting, baling, and storing hay. Hay bales in his day were the rectangular kind bound with wire, not like the huge round loaves lifted by mechanical grabs today. A strong man like Cleve had to use a hand hook to pitch the bale onto a truck and then off the truck into the barn. The job was hot, filled the lungs with powdery chaff, and required a strong back. A good mechanic was also necessary because the mower or the baling machine or the truck was likely to break down. Cleve could do every part of the work. He kept the tractors, mowers, and binding machine running. He was a good mechanic.

After Douglas's death and Rose's move to California, the little house in the Ranch House yard by the bois d'arc tree was vacant again. Cleve was asked to live there. He and his companion Dorothy helped with our barbecues, and Odessa, Emma Green, and Odessa's mother Narcissus were on hand for large occasions. Cleve and Odessa were the two people who watched over the aging Hawkins family members and helped them during the 1970s, when more frequent trips to Houston doctors became necessary.

I went to visit Odessa on December 1, 2000. It would prove our last meeting. Her house in Bay City was a cube, its roof a single peak in the exact center of the block. The house had one external adornment—shutters at the windows with half moons cut into their tops. The Hawkins sisters had had the house hauled into town for Odessa from Lizzie's place after her holdings were reincorporated into the ranch. The half moons on the shutters were a typical embellishment by the ever-embellishing Lizzie.

I was glad to find Odessa well. Her great-grandchild Dee climbed over her large lap. Odessa turned off the television, declaring, "What I like is some blessed silence." She told me about her own grandmother, whom she admired. Her grandmother had lived on an acre of land near Cedar Lane. "She had fruit trees and a garden and I used to love to visit her. She knew how to make a cake in the fireplace, an ash cake. She'd pour the batter right into the ashes that were just hot enough and not too hot." She could pop corn from the corn she grew, and she roasted peanuts, and every year Harry and Enoch would break her one acre of land and help her make a bale of cotton on it.

"She had an African name," Odessa said. "I wanted to be just like her."

Chapter 26

Frank Hawkins Lewis, Cattleman

The good people on the Hawkins Ranch were the almost daily companions of my brother Frank in every stage of his life. When he was about five, Dode Green supervised him in ranching occupations, letting him join the other hands in holding a rope that held down a cow. Later, when he was old enough to be a useful cowhand himself, he worked alongside Arthur Green, Harry Gatson, Tibby and Preston Wyche, Douglas Chapman, and Rose's father, Amos Franklin. Aunt Janie drilled Frank from when he was very young in the romance and family importance of the cattle business, and her instruction took such deep root that as an adult, he went eagerly and happily in the very direction toward which he was pointed. Aunt Janie took him into Houston to have the bootmaker at Stelzig's Saddlery fit him with blunt-toed black cowboy boots, and he had child-sized chaps made like those the ranch hands used. The man at Stelzig's fixed his black cowboy hat with a leather thong and slider to snug up under his chin.

While still in high school Frank went with my father to a livestock sale where he saw a pretty paint pony and fell in love with it. My father agreed to place a bid, and the pony came trotting out on a lead. To Frank's crushing disappointment, my father was absolutely silent, and the pony was sold. "You didn't say anything!" Frank said. "You just wait and see," said my father, who had gone around to consult Mr. Schwartz, a horse and mule trader who bought in large lots. As requested, Mr. Schwartz had placed the bid on the paint pony, and Frank brought him home. Frank named the paint Bobby and kept him in Aunt Janie's cow lot and shed. The first thing Frank taught Bobby was to rear up on cue like a horse in the movies. Frank saddled Bobby, rode him, worked cattle on him, curried him, and worked with him for hours to get him into the right gait; a pacing gait called the "single foot" was the ideal.

Often Frank got one or two of his high school pals, like Frank Montague or Gordon Richardson, to team up with him for ranch work, and the boys were paid for their efforts. With the regular ranch hands, they herded cows and calves to the loading chute for shipment, helped at the dipping vat, threw down the calves to be branded, marked, and castrated. They were not put off by heat, dust, blood, grease, or mosquitoes.

Sometimes the boys went on a coon hunt with John Ashcraft. He spoke in a high-pitched twang and loved to set loose his coon hounds at night and give them a holler to spur them on to catch some varmint in the woods. A night hunt with coon dogs was a favorite treat for my brother Frank. At our dinner table he described his expeditions, including a rendering of Ashcraft's dialogue about his dogs: "That's ole Blue," Frank said, pitching his voice a few notes higher than usual, like Ashcraft's. "He's got that weasel that's been getting at the hens." We asked ourselves how in the world Mr. Ashcraft knew from Ole Blue's howl that it was a weasel and not a squirrel or just empty baying at the moon.

The students in Bay City High School must have been required to read Thoreau's *Walden,* because one day at the dinner table Frank described Thoreau's cabin, and my father immediately gave him a challenge and a project. "Why don't you boys go down to the ranch and find a place on the Canoe Lake to build a cabin like Thoreau's?" It was summer. With lots of slack time, Frank and some friends took the challenge and went to the lumber yard for tools, boards, nails, tar paper, and shingles. Fed by Canoe Bayou, Canoe Lake is strangely shaped—on a map it looks like a hammer. They selected a spot with some shade under scrubby oak trees. The project kept them busy every day for the whole summer. My mother gave the bridge foursome regular reports on the construction project, all of them agreeing that an idle mind was the devil's workshop. At last the boys camped out in their cabin, which they called the Shack. The structure lasted three summers before collapsing in a heap.

While working on the Shack at the lake, the boys began to notice a lot of small turtles poking up their heads at the lake's surface. They hit upon the notion of catching the turtles and selling the little shells to one of the local merchants, so they went back to town to see the merchant and get a cast net. The merchant said sure, he could sell some of those little turtle shells if the boys supplied them. At the time, the fellows who wore neckerchiefs—cowmen, boy scouts, anybody who wanted a country look—were then sporting shellacked small turtle shells as fasteners for their neckerchiefs.

The boys waded into the lake and netted the turtles. My mother told the foursome: "Guess how they figured out to clean out the shells? They put them all on top of red ant beds and let the ants do the cleaning." In a few weeks' time, after the ants had done their work, the boys hauled their inventory into town, ready to transact business. They had bushels of turtle shells and dreamed of big money. But the merchant said no, he could not use bushels of shells; he was in the market for five or six. My father pronounced it a good lesson in economics: "You boys just broke the market in turtle shells. Supply and demand, you know."

On May 20, 1937, Frank was among the forty-eight members of the graduating class of Bay City High School. After parties, songs, a senior play, and the exchange of gifts—pens, combs, belts, wallets and sweaters—they marched across the stage in their caps and gowns and took their diplomas from the hand of E. O. Hutchinson, superintendent of schools. Mrs. Early was at the upright piano pounding out a march by Sir Edward Elgar. It was a final linking of arms for a tightly knit group before the breakup. For the rest of Frank's life, there was no moment when he did not love having a group of bosom buddies who knew the same talk and retold commonly remembered stories. Close pals provided him with insulation against the chafe of the culturally unfamiliar. Anything strange or foreign could provoke him to dismissive scoffing—a tenderfoot, a pretentious artist, or even anyone unconventionally dressed.

In aesthetic appreciation, Frank gravitated toward folk art and those skills that had been passed down from an older generation: the maker of saddles, harnesses, and boots; the skilled carpenter and creative iron worker; the expert at stitchery, weaving and quilting. Such artisans won his admiration, and he enjoyed elderly people from whom he might hear of their experience of the past. His interest was not compelled by world travel, museums, concerts, or theatrical performances. Frank's sense of identity was deeply planted in his understanding of southern culture, and he took the past and the local ways as his benchmark.

At the time of his high school graduation, Texas required only eleven years of schooling, and our parents thought Frank could use more preparation before going to college. They settled on Woodberry Forest in Virginia, and he agreed, with some misgivings, to study there before applying to college. He packed his Bay City High School football team sweater in his trunk, along with the clothes indicated on the school list. "Taking that team sweater was sure a mistake," he later told me. "They razzed me all over the place for bringing it, and my suit jacket had so much shoulder padding in it that the boys just stretched out their arms pretending to measure me and claimed I wasn't going to be able to pass through the door!"

He got a big family send-off at Houston's Union Station. When the huge train pulled into the station spewing its gasps of steam, the sight and metallic smell of the engine so palpably announcing his imminent setting forth, I found myself half sick with empathetic homesickness. Woodberry Forest was close to Orange, Virginia; getting there from Texas took two days and a night on the train or a day and two nights. Frank had never seen or visited the school as a prospective student, as is usual today, and neither had my parents or anyone else in the family. But when he arrived, already committed, he immediately loved the buildings and the beautiful Virginia hills. The work was difficult, but he was not unaccustomed to hard work, and he rose to the challenge. He played sports and made good friends. Fifty years later he wrote about his experience at the school: "I was captivated by its strong southern atmosphere, the majestic beauty of the place at the foot of the Blue Ridge Mountains, and won over completely by a host of friends whose southern background was very similar to mine."[1]

From Woodberry he went to Princeton and chose economics as a major. He went partly because his friends from Woodberry had elected to go there and partly because our family's admired banker friend, William Kirkland from Houston, had gone there and encouraged Frank to apply. For six years, from 1937 to 1943, Frank rode back and forth on the train for holidays and summer vacations. When he came home, his arrival was a great family occasion. He immediately got in touch with his local pals and went to work at the ranch; his face was as pale as tissue paper after winter in a colder climate.

In the late 1930s and early 1940s, we young movie-goers began seeing in newsreels booted troops of goose-stepping Nazis. Hitler's ranting speeches seemed absolutely demented, and Mussolini seemed too open to ridicule to be serious. But after the attack at Pearl Harbor on December 7, 1941, we young people began to collect metallic foil, save rubber bands, and cultivate Victory Gardens. For the young men at Princeton the war meant that their next steps were determined for them. Frank's class of 1943 would serve in the military either before graduation or directly upon graduating. Princeton's ROTC unit was a field artillery unit and Frank joined.

Late one afternoon in the spring of 1942, my father stretched out the pages of the *Houston Chronicle* and read that field artillery maneuvers were to be held in August in Louisiana. He thought it a golden opportunity for Frank to see what service in the field artillery really required. Somehow my father smoothed the way for Frank to serve without pay in the capacity of second lieutenant during the maneuvers, although at the time he had no commission but only ROTC officer training. Frank received a letter of commendation for his service performed from July 25 to September 1, 1942. Then and ever after, Frank had a natural taste for command and the use of authority. He could have been a fine career officer; but of course he already knew very well what he wanted to be. He was from childhood and ever after a cattleman.[2]

Our family received the engraved invitation to the Princeton graduation exercises scheduled for May 29, 1943. There was a wartime atmosphere, and we had been told not to count on service in the dining car. As we boarded, Aunt Sister came to the train and loaded us up with a neatly packed box of chicken salad sandwiches, each wrapped in wax paper, a box of fruit, and an impressively large box of her famous fudge.

At Princeton we walked over the grassy, tree-shaded campus. I was fifteen, and it was the first university I had ever seen. In strolling around we saw a scattering of now relaxed young men, exams finished, busy only with packing. The style of dress was wilted khaki pants, dress shirt with no tie, and scuffed white buck shoes. My father was shocked to spot a pint whiskey bottle in the back pocket of one of the students.

The graduation exercises under the trees at Princeton and the other celebrations were an absolute wonder to me. I had never seen a sight as inspiring as the radiant procession with colorful academic regalia and puffy Oxford hats. I felt I could see learning itself, see learned people, see a university making itself visible and inspiring through its panoply of symbols. By tradition the salutatorian's speech was delivered in Latin, and someone had quietly slipped into the hands of each graduating senior the printed Latin text with clues that said "applaud here" or "laugh here." It was an old Princeton graduation joke, and I was fooled. This bit of foul play did not diminish for me the beauty of the day; nor did it ever silence a self-addressed whisper that beckoned me to academia.

At dusk and deep evening after graduation, groups of seniors—all men in those days—walked shoulder to shoulder through the campus carrying lights and singing their wistful, sentimental school songs. The men's voices, harmonizing, seemed as full and deep as the "Pilgrim's Chorus." Soon they would go off to war. I, a fool for liturgy, thought it a shame that Frank and the other newly drafted seniors had to wear khaki uniforms to their graduation instead of cap and gown.

During his training at Fort Bragg, North Carolina, Frank had many opportunities on the weekends to visit our relatives, Jimmy and Lucy Brodie in Henderson, North Carolina. Jimmy remembered visiting the Ranch House in Texas when Frank Hawkins was still living; Jimmy's mother, Virginia Hawkins Brodie, was Frank Hawkins's sister. Now in semi-retirement, Jimmy ran a dairy farm, and that operation very much interested Frank, as did all agricultural projects. When Jimmy learned that Frank had won a saber for his outstanding ROTC record, he went straight up to the attic to fetch the Confederate Army saber worn by his father, Edmund G. Brodie, and presented it to Frank. Perhaps this Confederate saber had been worn at Appomattox; E. G. Brodie's unit was there at Lee's surrender.

After his training in field artillery specializing in long-range gunnery, Frank served in the Pacific Theater, primarily in the Philippines and Japan. He wrote home frequently. On February 11, 1945, from the Philippines he reported to his parents that he had received letters, Aunt Sister's photos, and the *Bay City Tribune.* He had not yet received the coffee pot and utensils that Aunt Janie had sent, apparently at his request. He said that three days ago his position had come under fire and that it was not "a comfortable feeling at all." Although the Japanese did not have as many guns as the United States, "they have been using what they have pretty steadily . . . and we have had some pretty close ones." He assured his family that no one had been hurt by the shells and that "there is almost always time to get under cover before they land because you can hear the whistle about 3 sec before they hit." He was pleased with the reaction of both officers and men in coming under fire for the first time. "All of them were careful to stay well under cover but there was no sign of fear or confusion."[3]

On Frank's return home from the war, he took over management of the Hawkins Ranch from Aunt Janie, as she had long planned. As soon as Rose Chapman learned that Frank held the rank of captain, she gave him the affectionate nickname "Cap'n," conveying not only the fact that he was now the boss but also that she had now bestowed her imprimatur.

Frank's management of the Hawkins Ranch was conservative. He did not embark on any radical or expensive changes. His aim was to continue in the pattern already set and to add only tested new practices found to be worthwhile.

His first effort was clerical. He would begin to keep the Hawkins Ranch books, which gave him a clear picture of the sources of income, the number of cattle and calves sold, and the major items of expense. My father, always ready with a suggestion, emphasized the importance of this clerical task. "If I were managing any kind of business, I would certainly want to keep the books. I would *want* to do it!" Probably our father's emphasis was to make certain that Frank took a businesslike approach to ranching without getting too wound up in riding horses and working cattle—the outdoor part of ranching he enjoyed. The emphasis was unnecessary; Frank took to heart anything our father advised. In an upstairs office on the Bay City square, Frank worked methodically on the ranch books.

On the ranch itself he set about improving fences and adding new ones as herd management indicated. He fenced off part of the pasture in front of the Ranch House for a hay meadow, where Cleve Woodward was a major hand. The native grass made good hay and was stored in the barns at the Ranch House. In time Frank also built several winter wind breaks and feeding stations far south of the Ranch House to protect and feed cattle that, for the most part, sustained themselves in both winter and summer by foraging in the open prairie. He was vigilant in taking care of windmills, sometimes having new water wells drilled, and seeing that the concrete troughs at each windmill were in good repair and the perpetually boggy places around the troughs were filled in. For a winter source of feed, he dug out a below-ground trench-type silo near the place where Lizzie had had her old headquarters. At that location there was a little rise, a small hill, which provided ideal drainage. With a tractor the hands dug a broad trench into the hill so that the floor of the silo could drain downhill. A sorghum crop, planted for the purpose, was cut while it was green and was packed into the pit by the stamping feet of the ranch hands. Over this heap of silage they added water to make it cure. Before the final covering of dirt, the hands put in several sheets of tin roofing. These firm lids allowed access to one segment of the silage at a time.

One readily acknowledged mistake Frank made was in trying to improve the blood lines of the Hawkins Ranch cattle. Although hardy range cattle, they were a haphazard mixture, varied in size and color and often brindled. They did not show well. The King Ranch had developed the beautiful Santa Gertrudis breed, which made a handsome herd uniformly red in color, and ranchers knew that uniformity of color and conformation was a marketing advantage. Frank thought the Hawkins Ranch herd would be improved if he could replace the rawboned look with a higher incidence of red color and a beefier conformation. He went to the cattle market and bought a few Hereford heifers and bulls to introduce a new blood line. The Herefords were beautiful with their red color and contrasting white faces. They had thick curly hair bred to withstand the cold blasts of wind in Britain; but on the Gulf Coast of Texas, once the summer heat bored into them, they panted and drooled. Even a short move from one pasture to the next prostrated them with heat exhaustion. Worse, their eyelids were susceptible to pink eye, which infected all of them. They were an almost total loss.

As a young bachelor back from the war, Frank found a brisk social life. He spent time with his old school friends, although some had scattered in these postwar years, and he attended a number of debutante balls in Houston. And when our cousin Martha Rugeley, daughter of Rowland and Daughty, married Dick Bachman in Bay City on January 20, 1948, Frank met her University of Texas roommate, who was an attendant in the wedding. She was Florence Neely of Amarillo.

Soon the news in the family was that Frank and Florence were to be married. All the family—parents, aunts, cousins—and many friends climbed into their automobiles and drove the six hundred miles from the Gulf Coast to Amarillo in the Texas Panhandle. We passed through fields of sunflowers, all with their faces turned the same way, and as we approached Amarillo, we began to see tall grain elevators on the horizon. This was wheat-growing country, and the Neely family were engaged in farming that crop. Florence's father had given her the childhood nickname, "Tootsie," which she hoped would disappear when she married and moved away. But it stuck, and she weathered it with good humor.

Romanticism put it into the heads of the bride and groom to make their home at the Hawkins Ranch House, the house that had stood since 1854, the homestead of J. B. and Ariella Hawkins and then of Frank and Elmore Hawkins and their children—our mother and her siblings. It still was standing thanks to the young lady ranchers having saved it in 1935. And it was unoccupied. But once Frank and Tootsie actually tried living in the 1854 house—by now with indoor plumbing but still no central heat or air conditioning, no telephone, no near neighbors, and no escape from mosquitoes—their original enthusiasm cooled. Soon they moved into town and bought a house. From town Frank could attend to his duties by driving the seventeen miles to the ranch, just as Aunt Janie had done.

His happy place of work remained a focus for leisure activities as well. For decades, Frank and Tootsie's recreational visits—bird shoots, barbecues, and cookouts near the Currie house—were filled with children, family, and friends who bridged the generations. In town Tootsie enjoyed the card table and continued the bridge foursome tradition, although occasionally the players turned heretical and played canasta or onze. Frank had no liking for cards or games; he said his biggest trouble with bridge was the "requirement of arranging the cards in that little fan."

Socially and professionally Frank and Tootsie became active in the Texas and Southwestern Cattle Raisers Association, which Frank served as director and president. Their attendance at association meetings meant they and their children got to know other ranchers and their children.

Frank and Tootsie lived out their lives among friends in the familiar town of Bay City, and each served on many civic projects. Over the years Frank enlarged his agricultural interests to include rice farming; in 1962 he became an officer of the First National Bank, which in time he served as chairman of the board. Into their lives came their four children, one after another from 1950 to 1956. In the years ahead they and my own four children would grow up together and take part in directing the activities of the Hawkins Ranch.

Frank Hawkins Lewis died on June 6, 2003; his wife's death preceded his on June 21, 1996. The acres of the Hawkins Ranch that came from J. B. and Ariella Hawkins, from Frank and Elmore Hawkins, and from our mother and her siblings, were the lodestone of my brother Frank's life.

Chapter 27

The Future of the Sense of Place

While my brother's daily occupations after World War II were centered on the ranch and in Bay City, mine stretched from the family center—to college, working in a Houston bank, graduate school and teaching, and then to sharing the life of my husband and our four children in places that work prescribed. Holidays brought our family together again and to the Ranch House. Aunt Sister told me that she never once, in all her life, spent Christmas away from home and family. After college when I showed signs of wanting to work eighty miles away in Houston, she sighed in deep disappointment. How could there be a satisfaction elsewhere greater than the satisfaction of home? The ties with one another that threaded through the same beloved place knitted together the generations and connected scattered cousins of the same generation.

For several years the operations of the Hawkins Ranch edged toward the perimeter of my attention. After undergraduate years at the University of Texas, I moved to Houston and found myself in a swirl of young professionals. I met Austen Furse. "Furse," he explained (and the explanation charmed me), "is that thorny bush in Thomas Hardy's *Return of the Native,* but Hardy spells it with a *z* or 'zed,' as the English say." He was a young lawyer, Yale English major, football player, and Air Force veteran of World War II; and he had grown up in a small Texas town, as I had. We married in 1955.

In an early venture, when our first-born daughter Janie was a toddler, we went to graduate school at Columbia University, where he earned a master of laws degree, and I began work toward a PhD in philosophy of religion. I had the delicious experience of being able to study under scholars I had known only through books—Reinhold Niebuhr, Paul Tillich, and John Herman Randall. At the end of the decade, I had my degree and four lovely children—Janie, Austen III, John, and Mary.

My husband and I, thus outfitted with advanced degrees, illogically moved from Houston to Bay City, where he practiced law and where such degrees were of little use and sometimes prompted a comment about why we would have bothered. The red-tile-roofed house on Avenue G, vacant since Aunt Janie's death in 1958, became the home of our young family.

Our move to Bay City about 1960 meant that our four children grew up closely involved with my brother's children—Frank Jr., Janet, Meta, and Jim. They were close in age and often shared pleasant weekends together at the ranch. Their fondness for one another as well as their energy and intelligence would be a great benefit to the operations of the Hawkins Ranch. They would all become ranch partners and, together with my brother and me, would function as a kind of board of directors in regularly scheduled partnership meetings that began in 1991.

Living with my family in Bay City, I was surprised, in the spring of 1970, to receive a telephone call from Rice University asking if I would consider commuting to Houston to teach an already announced course—the professor who was to have taught it had left to teach at Princeton. I began a fifteen-year part-time career as a university lecturer. In 1974, after some thirteen years in Bay City and three years of my commuting to Rice, we moved to Austin, where my husband became an assistant attorney general of Texas, and I taught in the Program of American Studies at the University of Texas until we both retired in the 1980s. Ritually for holidays my husband and I returned with our children to Bay City and to the Ranch House.[1]

Much more often than holidays afforded, I also returned to join in Hawkins Ranch business discussions as they arose. I was and am a partner in the Hawkins Ranch and also a trustee of ranch-owning trusts. Ranch business often called me to sessions with my brother and with Frank Jr., Janet, and Jim—those of his children who first became active in our ranching business and through many years made significant contributions to it. My part-time teaching schedule meant I could usually participate, and I also had time to write several books in my field of study. My ranching world and my academic world were so opaque to each other that, shuttling between them, I felt like an undercover agent with dual passports.

In 1991 those two worlds came closer together when I realized I had a teaching function that might be useful to our Hawkins Ranch family. Some of the children, now getting to be adults, did not have enough information about our common interests. I suggested that the family schedule a meeting and hold an orientation for them. Because teaching had given me experience in choosing and preparing material for students, I undertook to create a one-day seminar. With considerable help from our ranch office, I prepared a large notebook for each family member (ten of us), and on April 27, 1991, we all met. There were many lasting benefits of that session, of which I focus here on two.

The first was that we all became clear, for almost the first time, about what our main business really was. Data tracing average income for three years showed us that our main business was an investment portfolio plus mineral interests. The answer surprised us a little, because much of the hum and bustle of our ranch office had more to do with the costs of mowing pastures, repairing fences, baling hay, dealing with rice farmers, replacing culverts, repairing windmills, and other farming and ranching details.

The second benefit was that we began to make the annual meetings routine, and we now meet quarterly, follow an agenda, and track the status of the items discussed. In this and many aspects of our Hawkins Ranch partnership, we have been especially aided by Bill Isaacson, whom we recruited as business manager in 2003.

Because we now have new clarity in the data, we have been able to see what changes we might choose to make. We made a surprising change in 1999. When the data of our ranching operations were assembled and presented to the partners, we saw that by then our business of raising cattle broke even only every other year. One heard of ranch owners who with the help of mineral production on their land played at wearing a hat and boots and herding cattle apart from profit motive. But the Hawkins Ranch partners wanted to be businesslike and, in any case, we had little mineral production located on the Hawkins Ranch land itself.

On April 1, 1999, the partners sold their Hawkins Ranch herd of cattle to a tenant, who then grazed the same herd on Hawkins Ranch land. This sale constituted an immense swerve from the inherited course, ending a family cattle business that had operated continuously for 129 years. The end came with understandable wistfulness but no indecision, because the partners had before them clear financial data on which to base their decision.

The sale was one of the reasons that the main business of the Hawkins Ranch in recent years has become more concerned with a diversity of land, mineral, and investment interests and less with crops and cattle. Ironically, some of the proceeds of the sale went to make the Hawkins Ranch House more comfortable and useful for the current family, who were careful to maintain its historical footprint as J. B. Hawkins had built it in 1854.

There is, however, another and even greater reason for the recent trend toward diversification of holdings. It derives from the decades of business activity of J. C. Lewis and Esker McDonald, the two banker brothers-in-law. Their land acquisitions and their farming, mineral, and banking interests were not initially a part of the Hawkins Ranch family business but became part of it through their wills or through trusts they created. From the 1940s to the 1970s, the two men had been in partnership and had bought land on which they successfully farmed rice and grazed cattle in rotation with planting. Their years of success increased their respective holdings in a portfolio of securities as well as in land. During this period, two of their land acquisitions, the River Ranch and the Buckeye Ranch, developed significant mineral production, a fact that again increased their holdings in marketable securities.[2]

On the acres of the Hawkins Ranch itself, many mineral leases have been made and have been a source of revenue, but there has been no significant production on these acres that matches that of the River Ranch or Buckeye. The earliest prospect on the Hawkins Ranch was in the 1930s, when B. C. Cockburn drilled a well in the Sheppard Mott. It was a time when drillers in the county looked over land and by its look or by divine instruction (as was the method of Edgar B. Davis, an early operator at Buckeye) declared that a well should be located "here."

I remember Cockburn's well location. Sometime in the 1930s I was with my parents on the Sheppard Mott road when rain began converting it into a series of muddy wallows. Several family cars were in a line, returning home after an open-air supper in the yard of the Ranch House while the evening was still clear. Esker's lead car gunned through the first mud hole, but the next car bogged down to the axle and blocked the way for those following. Esker offered to take as many people as he could to town and return for the rest of us. The night turned cold while we waited, within sight of the lights on Cockburn's drilling rig. "Let's go over there, they'll have a fire," someone suggested. The derrick crew had a big boiler and good shelter, and Daughty got a welcome tin cup of hot coffee handed to her. There was no marketable production from this well, but a small medicine bottle of dark oil was given to my brother and stayed on Aunt Janie's mantel for a year or two. "Well, Frank," said one of my uncles, "I guess you are taking your royalty in kind."

Still another factor leading to the diversification of Hawkins Ranch assets was the local bank stock that Esker McDonald and J. C. Lewis had acquired during their years of service as bank officers. This stock became greater in value in 1995 when it was bought by a large regional bank and later became a Wells Fargo bank.

A first step by which many of the assets of Lewis and McDonald eventually came into holdings of the Hawkins Ranch was through the creation of several trusts. The trusts themselves then became partners in the Hawkins Ranch. A second step occurred on December 23, 1994, with the creation of the Hawkins Ranch Ltd. partnership agreement. It furnished a big receptacle into which were placed almost all significant family properties, whether held by trusts or individuals. Each person or trust contributing properties to the partnership received a percentage of partnership interest. Ten of us became partners in the new Hawkins Ranch Ltd., as prescribed by our written document. My brother Frank Lewis and I were signatories as individuals and trustees, and the other signatories were our respective children.[3]

The Lewis and McDonald business activities increased the Hawkins Ranch portfolio of marketable securities, and their land holdings alone increased the acres of the Hawkins Ranch by about ten thousand acres. This infusion and diversification means that the Hawkins Ranch, if understood in terms of its assets, is no longer only a *place* with grazing cattle. Once it could be defined with a wave of an arm toward those contiguous acres generally bordered by Liveoak Creek, Peyton Creek, and Lake Austin—those "broad prairies" of which Sallie Hawkins wrote, those acres of the Duke League surrounding the Ranch House that provoked Ariella to assert her community interest. The Hawkins Ranch Ltd. still owns these beloved original acres, and the Hawkins Ranch House is still the iconic place of home, land, and kin. But the partnership has come to own land more distant from this house and these acres—land and investments acquired for utilitarian purpose and, therefore, with less power to compel family attachment.

This circumstance means that in decision-making, current partners must steer between the poles of heritage and business.

On the one hand, today's partners, having a diversity of assets to manage, must operate their holdings in a pragmatic, emotionally detached way—as a business. Doing so requires seeking new opportunities, taking on risk, and welcoming change. But change might lead to disaffiliation from the beloved place, for the place itself could one day become a practical opportunity, one that promises a present value exceeding its value as a received heritage.

On the other hand, partners are also family members and share deep affection for their long-held land, for their special Ranch House, and for the family bond that is strengthened by the beloved place. They view their heritage as a value received from the past and want to safeguard it. Change can be seen as betrayal, a sell-out, instead of a practical adjustment to circumstance. The tension between these poles, *business vs. heritage,* seems inherent in a long-term family business, and this tension sets the conditions for the current management of the Hawkins Ranch.

A few years ago, at Christmas time, I climbed to the bedroom on the south wing of the Ranch House, the room we call Meta's Room. It had also been the bedroom of Ariella and J. B. Hawkins. From there I looked out at the bois d'arc tree by Dode's house and at Lake Austin and beyond. I was puzzling over a decision that partners might have to make in the future. In exchange for a handsome royalty payment, would we permit an army of huge wind turbines to be cemented deep into the ground, across Sallie Hawkins's "broad prairies"? Our regard for heritage would turn away such a thing as a wind farm if it ruined the beloved place. On the other hand, our impulse toward practical business would at least make us weigh the prospect as an opportunity. We needed to know whether a wind farm would in fact be ruinous. It would be three miles away from the house, but what would it look like?

I began to think. It would help us decide if we knew beforehand how intrusive and menacing these giants would seem. Maybe somebody who knows trigonometry could figure out from measurements how high they would appear to rise above the horizon when viewed from our Ranch House. But who could do that? In musing about the question, my mind swept back to the alley way of the town and to my childhood pal, the boy who knew everything about the Morse code, chemical reactions, and gravity. What we need, I thought, is somebody like Adelbert; he would know how to do it.

We Hawkins Ranch partners did not have an Adelbert to consult, but we did go on a fact-finding trip to already installed Texas wind farms. We plainly saw that a wind farm is an industrial site and a vast installation that overwhelms any other use or appearance of the land.[4]

We partners returned from viewing wind farms with the same feelings: that the inviolable center of our heritage was the Hawkins Ranch House itself. We thought that if there ever were to be an industrial site, it had better be as far removed from the Ranch House as possible, preferably not on the original contiguous thirty thousand acres of the Hawkins Ranch that came from J. B. Hawkins, his son Frank, and Frank's children. To the current partners these seemed to be heritage-bound acres, not to be treated in a purely utilitarian way.

This *heritage* vs. *utility* dilemma was apparent in the 1935 deliberations of the young lady ranchers about saving their Ranch House or letting it fall. They asked themselves if the value of the Ranch House as a heritage was worth the practical cost of saving it. The current ranch partners might very well be businesslike and opportunistic when it comes to impersonal holdings like securities or land many miles away from the Hawkins Ranch House. But with respect to the Ranch House and its immediately surrounding pastures, they would all still lean toward the heritage side of the dilemma.

The day may come when Hawkins Ranch partners have long been scattered and find themselves with no interest in sharing the experience of the special place with kin and partners whom they hardly know. They will have become unbound by the place, disaffiliated from one another, and then they will value it only opportunistically in purely business terms. So far that has not happened.

The earliest Hawkinses have left the current family a *place,* a material artifact that dates from antebellum times. The place that continues to engage family affections includes certain acres of the Texas coastal region, the familiar town, and especially the Hawkins Ranch House itself. It represents a visible sign of an interior thread that connects five generations. It also binds together members of the current generation, because they are pledged to good citizenship in a 162-year family partnership that is a business but also the custodian of a cherished place. "I think we will make a pretty place of it," were the words of James B. Hawkins while building the Hawkins Ranch House in 1854, and today the generations who enter its gate find themselves with the appreciative reply—and you certainly did.

Miss Tenie Holmes viewing pictures in the living room of Meta and Jim Lewis in 1945. Photograph by Margaret Lewis Furse, HRLTD

Ranch Hands at the House Pens in the 1940s. Mounted are Arthur Green on Socks (left) and Harry Gatson. Front, left to right: Amos Franklin, father of Rose, William, and Jim Franklin; Douglas Chapman, Rose's husband; and Pitchy Mills. Dode's House is visible at far left. Photograph by Margaret Lewis Furse, HRLTD

Cleveland (Cleve) Woodward, 1999. Photo by Beryl (Sis) Sprouse Cochran, HRLTD

Odessa Gatson Brown with the Lewis and Furse children, 1963. On her arm is Austin Furse III, at left is Meta Lewis, and at right are Janie Furse and Jim Lewis. Photograph by Margaret Lewis Furse, HRLTD

Rose Franklin Chapman with John Furse, November 1964. Photograph by Margaret Lewis Furse, HRLTD

Dode Green and ranch hands supervise children in holding a calf needing treatment, 1926. Dode is at far left in chaps at the fence. The children in foreground are Beryl (Sis) Sprouse (left) and her brother Claude (Buddy) Sprouse, and behind them is Frank Lewis, aged six. At the right is Albert Wadsworth. Photograph from the Sprouse Family Snap Shots, HRLTD

Frank Lewis (right) and his high school pal Frank Montague captured a bobcat, 1936. HRLTD

Picnics in the country became frequent during the 1950s with the marriage of Frank Lewis and Florence (Tootsie) Neely. Left to right: Meta in headscarf, Sister, Frank, Tootsie, and Esker McDonald. Photograph by Margaret Lewis Furse, HRLTD

Frank Hawkins Lewis with his Tennessee Walking Horse, 1984. Photograph by Patsy Wiginton, courtesy of Janet Lewis Peden, HRLTD

Jim Lewis in a rice field, 1950. Raising rice was central to the partnership of Jim Lewis and Esker McDonald. HRLTD

A Hawkins Ranch business meeting, November 7, 2008. Standing is Business Manager Bill Isaacson, and at far left is John Hawkins Peden, representing "generation six." Starting beside him, current partners are (left to right): James Neely Lewis, Mary Elmore Furse, John Lewis Furse, Margaret Lewis Furse, Janet Lewis Peden, Jane Hawkins Furse, Austen Henry Furse III, Meta Lewis Hausser, and Frank Hawkins Lewis Jr. Photograph by J. R. Mullen, HRLTD

Hawkins Ranch partners investigate wind turbines near Midland, October 16, 2008. Left to right: Jim Lewis, Margaret Furse, Mary Furse, and Janet Peden. HRLTD

Acknowledgments

In writing this book I have had the guidance in early stages of Howard Lamar, William H. Goetzmann, and Randolph Campbell; and of friends who are themselves writers and editors—Anita Howard, Kathleen Niendorff, and Ann Friou. I have also found that the twenty happy years of discussing books with book club friends have helped me catch sight of those factors that chance to make an imperfect book a better one.

For their interest, hands-on help, and encouragement my thanks also go to my children Jane Furse, Austen Furse III, John Furse, and Mary Furse as well as to my nephews and nieces Frank H. Lewis Jr., Janet Lewis Peden, Meta Lewis Hausser, and James N. Lewis. I have noted in my introduction that my granddaughter Elizabeth Friedman and son John Furse joined me in searching through the Hawkins and Alston letters at the University of North Carolina. Jane Furse and Anne Seel Furse, both good writers and editors, gave me helpful advice on organization.

Wayne Bell, architect, whose field is historical site preservation, planned the 1999 renovation of the Hawkins Ranch House—work finished by the architect Logic Tobola. Both improved the comfort of the house while preserving its historical integrity. The architect James Robert Coote kindly read my description of the Hawkins Ranch House and saved me from many, though probably not all, blunders.

My cousin Martha Rugeley Bachman helped identify family photographs; Helen Cates Neary set me straight on details of the county courthouse; and Paul Hemmer pointed out the census report giving evidence of the use of prisoners on the plantation. Helping me fill in other local details were Frances Parker, Mary Belle Ingram, Carol Sue Gibbs, and Virginia Moore Whiddon (the childhood friend who joined me in exploring the town).

Until his death in 2003, my brother Frank H. Lewis took care of historical records of the Hawkins Ranch in a metal World War II shell box. His preserving the records has been a practical help. Greater yet, however, has been his communication to younger generations his own regard for the Hawkins Ranch as a beloved place.

I especially thank the office staff of Hawkins Ranch Ltd.—Bill Isaacson, Lena Stavinoha, and Cindy Tomek—for organizing and retrieving many local and ranch records, maps, and photographs. Lena Stavinoha is to be thanked for creating and organizing the Hawkins Ranch File of Historical Data, a source frequently referenced in this book.

John Robert Mullen and Patsy Wiginton were the photographers for several handsome illustrations used. Bill Isaacson created maps of the Hawkins Ranch especially for this book.

Institutions I would like to thank are the Matagorda County Museum of Bay City, Texas, and the Wilson Library of the University of North Carolina, which holds in its Southern Historical Collection the Archibald D. Alston Papers and the Hawkins Family Papers.

Appendix

SKETCHES AND LETTERS OF THE ANTEBELLUM CHILDREN

The Hawkins family story in Texas follows the lineage from J. B. Hawkins through his first Texas-born son, Frank Hawkins, the father of my mother and her siblings—Harry, Meta (my mother), Janie, Lizzie, and Elmore (called Sister). In the family story I allow the brothers and sisters of Frank Hawkins to drop from sight, but as children they figure in the letters their mother Ariella Hawkins wrote from the plantation. Following are brief identifying paragraphs about the nine children of Ariella and James B. Hawkins, listed in birth order.

CHILDREN OF ARIELLA AND JAMES B. HAWKINS

Sallie (or Sally, 1837–60) was the eldest child. She was named for her grandmother, Sarah (Sallie) Madaline Potts Alston. Sallie Hawkins was much loved in the family and looked to as a friend and confidante by her younger sister Virginia ("Little Sis"). Sallie's first attempt as an aspiring artist was a painting of her father's sugar house.

Sallie attended a school in Warrenton, North Carolina, and then Patapsco Female Academy in Maryland, where among other subjects she studied botany and art. The headmistress of Patapsco, Almira Hart Lincoln Phelps, the younger sister of Emma Willard, established an unusually ambitious curriculum for young ladies in that day. Mrs. Phelps was a noted scientist and the author of textbooks in chemistry, biology, botany, physics, and geology for secondary schools and colleges. Sallie studied botany under her and used her botany text in class.[1]

The full title of Sallie's copy of the text (from the Hawkins Ranch File of Historical Data) is *Familiar Lectures on Botany; Explaining the Structure,*

Classification and Uses of Plants, Illustrated upon the Linnaean and Natural Methods with a Flora for Practical Botanists. For Use of Colleges, Schools, & Private Students. By Mrs. Almira H. Lincoln (now Mrs. Lincoln Phelps), Principal of the Patapsco Female Institute of Maryland; author of "The Fireside Friend," a series of works on Botany, Chemistry, Natural Philosophy, Geology etc. (new edition, revised & enlarged. N.Y.: F. J. Huntington and Mason Brother 1854. First published 1829).

On December 8, 1858, in the town of Matagorda, Sallie married Ferdinando Stith of Memphis, nicknamed "Buggie." Her untimely death at age twenty-three occurred in Memphis and was probably caused by yellow fever. She is buried at the Hawkins plantation family cemetery in a vault that her father had constructed after he himself went to Memphis to recover her remains for burial at the plantation. Virginia said bringing her "home" was a consolation. She had no children.

Willis Alston Hawkins (1839–63) is named for his mother's distinguished father Willis Alston, planter and member of Congress. (Willis Hawkins's first cousin Willis Alston, son of Ariella's sister, is also named for him.) Willis Hawkins went to school in Warrenton, North Carolina, and on November 3, 1857, married a young woman of Warrenton, Leah T. Irwin. In 1858 he and Leah prepared to set up their own plantation next to his father's on Caney. His uncle, John D. Hawkins Jr., his father's partner, said he wished "to aid Willis in settling his matters here [in North Carolina] and take his Negroes to Texas."[2] The census of 1860 lists Willis and Leah as residents of Warrenton and also of Caney, a duplication of their names and properties, including the same number of slaves, eighteen.

On June 23, 1858, J. B. Hawkins had purchased from Major Prewitt 2,028 acres of land in the Prewitt League, including 1,000 acres that J. B. Hawkins then conveyed to Willis by deed on February 28, 1859. The 1,000 acres, described as being on the west side of Caney Creek, were sold on November 28, 1861, to Asa W. Thompson. The speed with which Willis acquired and then sold his acreage on Caney can possibly be explained by his imminent service in the Confederate Army.

Further evidence of their living in North Carolina comes from Virginia's letter of November 18, 1860, to her grandmother at the time of the family's grief over Sallie's death (see letter following). She tells her grandmother that "Bunlley [a nickname for Willis?] and Leah have been with us for two weeks. You don't know how happy I was when they came. I never want them to leave us again."[3]

Willis served in the Confederate Army, having been commissioned a captain in Company A of the 12th Infantry Regiment Georgia and promoted to lieutenant colonel in 1863. On January 24, 1863, he resigned, perhaps because of illness, and he died that same year of unknown causes. He is buried at Oakwood Cemetery in Raleigh, North Carolina. The grave marker only indicates his initials.

Leah lived as a widow in North Carolina, and while helping her pregnant sister keep house, wrote to Willis's grandmother, Mrs. Alston. The letter was written on March 26 from "Fair Mount," but she does not give the year; it was probably 1863 or 1864, following her husband's death, as she does not mention him, and certainly she would have if he had been living. Leah's letter is one of the two examined that gives a picture of the pinched conditions in Warrenton and Raleigh during wartime and the growing sense of the failure of the Confederate cause.

Leah continued living in North Carolina and married Joseph E. Drake, who in 1889 joined her in selling the remainder of the land (1000 acres) that she and her late husband, Willis Hawkins, had continued to own west of Caney Creek. Judging from this second sale of land, J. B. Hawkins must have conveyed to Willis in two parcels of about 1,000 acres each the 2,028 acres in the Prewitt League that he purchased from Major Prewitt in 1858.

Virginia (Jenny, Jennie, or Jeannie, 1840–1930), like her sister Sallie, went to school in Warrenton and then to Patapsco Female Institute in Maryland. On February 7, 1877, she married Edmund Gholson Brodie of Henderson, North Carolina, where she lived all the years of her long life. The couple had one child, James Hawkins Brodie, the first cousin of my mother and her siblings, who always kept in close touch with their well-liked cousin "Jimmy Brodie." As an adult he became co-trustee with Henry Rugeley of the five Hawkins children's estate while they were still minors. Jimmy's father, E. G. Brodie, as a new graduate of the University of North Carolina in 1861, wrote a letter to the Alston family in North Carolina describing local preparations around Raleigh for the pending Civil War and his own imminent enlistment in the Confederate Army.

James Boyd Hawkins Jr. (Jimmy, 1843–57) died at age fourteen and is buried at the Hawkins family cemetery at the plantation on Caney, Hawkinsville, Texas.

John Davis Hawkins II (1844–65) is named for his grandfather. He is one of three John Davis Hawkinses: J.B.'s father, J.B.'s brother and partner, and this son. He grew up on Caney and served in the Civil War in the 35th Texas Cavalry, Company D. He died at age twenty-one and is buried at the Hawkins family cemetery at Hawkinsville, Texas.

Ariella (Ella, 1846–51), described by her mother Ariella as the "sweetest child," died when she was only five and is buried in the Hawkins family cemetery at Hawkinsville, Texas.

Frank Hawkins (1847–1901) was born on December 17, 1847, in Matagorda in the Christmas season, when Ariella was "quite alone" but "never did better." He was the father of the five Hawkinses born at the Hawkins Ranch House and highlighted in the chapters of this book. In 1866 he, alone of the children of J. B. Hawkins, was sent to Germany and then to England for his education. In the 1870s he was responsible for changing the family business from sugar to cattle. He and his wife, Elmore Rugeley, lived at the Hawkins Ranch House, not at the Caney plantation. Frank Hawkins is treated in chapter 8.

Edgar (or Charles Edgar, 1849–87) was named for two of Ariella's brothers but was always called Edgar. He married Ann (or Annie) Lewis Hardeman and the couple had four children: James, Frank, Edgar (called Ned), and Ella. Edgar and his family lived at the Caney plantation long after sugar production had ceased; he died at the age of thirty-eight and is buried in the plantation family cemetery. Edgar's children and the children of Frank Hawkins were close in age and in the feeling of kinship. Edgar's family inherited the Caney plantation house and property. The Frank Hawkins children inherited the Hawkins Ranch House and lands.

Annie died as an infant, two months of age, and is buried in the Hawkins family cemetery at Caney. Her grave marker indicates her date of death as 1852. She is not mentioned in letters.

A Letter about the Death of Sallie Hawkins

Sallie Hawkins, married to Ferdinando Stith and living in Memphis, fell victim to some epidemic there, probably yellow fever. At the peak of her happiness, she died. Her sister Virginia, devastated by the loss, wrote to their grandmother Sarah Alston of the terrible sadness:

> Magnolia Caney
> Nov 18, 1860
> My dear Grandma—
> How different are my feelings now and when I wrote you last. I was then hopeful and happy expecting dear Sallie to be with us in Feb. Her last words to me, on bidding me good-by were 'Little Sis I am coming home in Feb.' The day I wrote to you I wrote her a long letter. She had been dead then a week. We did not hear it in nearly three weeks owing to the mail. But—oh! Grandma, I cannot think of her as being dead, her whom we all loved so dearly and the last one I thought of being taken from us. OH! It does seem so hard. Pa received a letter from Buggie [Sallie's husband Ferdinando Stith] the other day. Poor fellow, I know what he must suffer. He was not with her. Pa is going on after Christmas when the weather gets cool to bring her home and put her by the side of Jimmy, he is going to have a vault built. Buggie wrote that he had her remains placed in a private vault in Memphis where it would remain until Pa went on. We will all feel happier to have her here. Bunlley [a nickname for Willis?] and Leah have been with us two weeks. You don't know how happy I was when they came. I never want them to leave us again. I wish dear little Johnny could come home and stay. I shudder when ever the mail is opened fearful of hearing some dreadful news. We are now staying at the plantation. I don't know how long we will remain. We came down to be with Pa, he is obliged to remain here during tolling season. I believe his crops both cotton & sugar is good. The weather is still very warm, almost like summer. I will write again soon, Grandma. Best love to all, Very affectionately
> Jeannie[4]

The reason "dear little Johnny" (John Davis Hawkins II) was absent at the time of the family's bereavement is not known; aged sixteen, he was perhaps away at school. There is no confirmation that "Bunlley" was a nickname for Willis Hawkins, Leah's husband.

Two Civil War Letters

Following is a letter from Leah (Mrs. Willis Hawkins) to Sarah Alston describing the domestic effect of the Civil War in North Carolina. The year of the letter is not given but as noted earlier it was probably after 1863, the year of Willis's death.

Fair Mount
March 26th
Dear Grand Ma

It has been so long a time since I have seen or herd [*sic*] from you that I have concluded to write and find out your "whereabouts," and what you have been doing with yourself all this time, as it will be impossible for me to come and see you soon. Sister *expects* in a few weeks, and I will have to stay at home to keep house, and not a very pleasant employment these times, such a hard matter to get anything *good* to eat.

Grand Ma, what do you think of the war now? Things begin to look pretty gloomy, don't they? But tis said the "darkest hour is just before day," and I don't think times can be much darker for us, unless we have *all night*—and I have no idea of subjugation. I think we will have to endure many hardships, but will finally gain our independence.

I do wish so much I could get to Texas. I am more than anxious to be with them all—but I am afraid it will be impossible as long as the war lasts. I haven't heard from them since Nov—had a letter from Ma dated the 8th and intended sending it to you, but heard you had received one later. I sent word to Aunt Jane that all were well & getting on finely [illegible phrase].

I spent a few days in Raleigh about a month ago. Saw Charly there but I suppose he has been at home since then. I thought he was looking as well as I ever saw him and seemed to be enjoying himself very much *riding around* and I reckon carrying on with the girls. I went to do a little shopping in Raleigh, found plenty of goods, but most of them beyond my purse. Alpaca was a hundred dollars a yard. Bleaching $25 and nearly everything else in proportion. And I concluded I could do better in Warrenton.

Grand Ma you must write before long. I don't know whether you are with Aunt Zuri or Mary, but will send this to Warrenton and if you are not with Aunt Zuri, she will send down to you. Write me about the children, where Bettie is going to school and if Willis is still at Chapel Hill. What has become of Miss Dickens?

Give my best love to all.
Hoping to hear from you soon.
Affectionately, your granddaughter,
Leah Hawkins[5]

E. G. Brodie, who would later marry Virginia (Jenny) Hawkins, wrote to a friend describing war preparations.

Henderson NC
May 16, 1861
Dear Friend

Only a few days have elapsed since I received your letter although it had been written some time when I got it on account of its being delayed in the office at the Hill [the University of North Carolina at Chapel Hill]. I have been home some time got weary of the Hill spoke on Monday and left the same night with permission, got my bible and an assurance of my diploma.

I joined a company in Henderson and it was so long getting ready for battle, I left them and am now [with the?] Greys stationed in Wilton [word unclear]. We are in the second Regiment. I think also another company from this county and 2 from Warren. Everything is war and what pertains to it with people of this state. I was in Raleigh Tuesday, it is literally a tent and drill ground for soldiers. Our Col. (Sol Williams) is in Raleigh it is said for the purpose of getting the Gov's permission to quarter in the fair grounds here—Military fare and discipline is not very desirable unless everything is fixed better than we have it but everybody is cheerful. I think perhaps we have the nicest company I have yet seen candidly. It is composed of the young men only of Granville. There are only a few in college at this time everybody from the county has left except Arch Henderson. I am only to stay here until Saturday and my furlough is out. Venerable Hargrove & Royster were elected to the Convention in this county & Arch Williams in Franklin. I cannot say where we will be so that you may direct your letters as we are expecting to move every day either to Norfolk or Harpers Ferry. Direct your letters to Henderson & Pa will forward them. Stewart is married to Mss McIntosh.

Your true friend,
E. G. Brodie[6]

Notes

INTRODUCTION

1. In the memorial address delivered at Warrenton, N.C., on September 29, 1829, commemorating his late grandfather, John Davis Hawkins (father of the first Texan, James Boyd Hawkins) said in part:

> There were other branches from the Charles City stock, [Charles City County, Virginia, the birthplace of the memorialized grandfather] which migrated to other parts of the Union; one went to the state of Kentucky, which produced Joseph Hawkins, formerly a member of Congress from Kentucky, and who afterwards died in New Orleans. That gentleman traced his connexion with our family in a conversation with our distinguished and venerable fellow citizen Nathaniel Macon, Esq., who now contributes his presence to commemorate this occasion and this day." John D. Hawkins, "Colonel Philemon Hawkins, Sr."

See Cantrell, "The Partnership of Stephen F. Austin and Joseph H. Hawkins." Stephen F. Austin met Joseph H. Hawkins in New Orleans, where Hawkins was in the practice of law. Austin may have known Joseph's brother, Littleberry, while attending Transylvania University in Lexington, Kentucky, in 1808–10. As a young man down on his luck, Austin came to New Orleans, a promising commercial center, in hopes of finding almost any job. Hawkins took him in to train as a lawyer, lent him money, had Austin take his meals with his family, and eventually joined him as a partner when Austin assumed his father's colonizing efforts. As documented by Cantrell, Austin was scrupulous in the division of his land grant with the heirs of his then deceased partner, Joseph H. Hawkins. The Joseph H. Hawkins children were George Nicholas Hawkins (the eldest son), Edmund St. John Hawkins (first to come to Texas as Austin's colonist), Mary Jane Hawkins (wife of William Victor, who questioned the fairness of the land division), Norbourne Hawkins, and Joseph Thomas Hawkins. The estates of Norbourne, Edmund, and George were all probated in Brazoria County in 1837. Joseph Thomas Hawkins lived until 1850 as a planter in Brazoria County. A relative but not a son of Joseph H. Hawkins was Major John T. Hawkins, a settler in Austin's colony.

2. Two collections of antebellum letters were used: the Hawkins Family Papers, #322, and the Archibald D. Alston Papers, #16, both of which are in the Southern Historical Collection of the Wilson Library at the University of North Carolina at Chapel Hill. The Alston Papers are the main source of letters quoted. They describe the 1846 beginning of the Hawkins sugar plantation in Texas and the building of the Hawkins Ranch House in 1854. Most are letters from Ariella Alston Hawkins in Texas to her mother, Mrs. Sallie Alston, in North Carolina. Other letters to Mrs. Alston from Texas are by James B. Hawkins and his daughters, Sallie and Virginia; Virginia wrote from the plantation house on Caney Creek and from the Hawkins Ranch House on Lake Austin. The Hawkins Papers consist of 28.5 linear feet of documents, most concerning the Hawkinses who stayed in North Carolina; but the collection includes letters James B. Hawkins sent from Texas to his father and brothers and letters from his brother Frank, who traveled with James to Mississippi.

3. James B. Hawkins to Sarah Alston, January 12, 1854, Archibald D. Alston Papers.

4. I was impressed by how well Ariella and the children of the plantation understood the family's livelihood and had a literal "view" of it. On the plantation there was no separation between home life and work life. The father of the family made plans for his crops, and for the building of his sugar mill (counting every brick), and his wife and children were as fully informed as he. For wife and children there was no mystery about what the father did for a living. They could look at the fields and see his living. Benjamin Franklin's father took him for a walk along the streets of Boston in order to see the kind of work from which he might choose a livelihood more to his liking than his father's candle making and fat rendering business; see Schneider, *Autobiography of Benjamin Franklin,* 12–13. Earlier agrarian and artisan work is much more visible than much of the work undertaken in offices today, even in farming and ranching offices.

Social change has made a separation between work life and home life that obscures the work of the family breadwinner for those not directly involved. This change is apparent in the history of the Hawkins Ranch, as can be seen in the final chapter. In the 1990s ranch business data, stored in electronic files, were accurate and complete but not easily accessible to partners. Instituting clearer presentations of data, coupled with regular partners' meetings, have been a help in demystifying the business facts in the 1990s.

5. Salamon, *Prairie Patrimony,* ch. 4, "A Typology of Family Farming Patterns," 91–116.

Part I: Plantation Beginnings, 1846

1. Marr, *History of Matagorda County,* 96–97.

Chapter 1: North Carolina Roots

1. John Davis Hawkins to John Branch, October 17, 1829, Hawkins Family Papers.

2. John D. Hawkins, "Colonel Philemon Hawkins, Sr." an address delivered by John D. Hawkins to his family in Warren County, North Carolina, September 29, 1829; privately printed in 1829 and 1906, it was later republished in the *North Carolina Booklet.* A copy of the address is archived in Hawkins Ranch, File of Historical Data, Hawkins Ranch Ltd., 2020 Avenue G, Bay City, Texas 77414.

John D. Hawkins declared: "Philemon Hawkins the second was born in Virginia to the first Hawkins (Philemon the first) to emigrate from England. The mother of the young son was widowed and then married again, but this second husband was a cruel man who abused her." John D. Hawkins then unfurled the edifying story of this young man, Philemon the second, his grandfather: "It was [the] misfortune [of Philemon's mother] not to find in her husband that conjugal tenderness, affection and forbearance, which the wedded estate should assure to those who enter into it." She was so mistreated by her second husband that her son's presence was necessary for her safety. Meanwhile, the young Philemon's ability and energy had caught the eye of a Colonel Lightfoot of Williamsburg, Virginia, "a gentleman of great wealth and discernment." Lightfoot had three plantations and proposed placing these estates under the young Philemon's management, and Philemon had agreed and signed a contract to that effect.

Then Philemon's mother's abusive husband proposed moving her and her young daughter, Ann, to North Carolina. But she was simply afraid to move away from her son's protection and risk mistreatment at her husband's hands. She pleaded with her son to break his contract so that he could go with them to North Carolina. Her son, ambitious to succeed in a good opportunity and feeling the obligation to fulfill his contract, refused his mother's request. Desperate, she went herself to Williamsburg to ask for her son to be let off from his contract, but she was too ashamed to give the real reason, and Colonel Lightfoot refused to break the contract. Rejected in her proposal, she spent a miserable night at a neighbor's home. "There melancholy, with all its accompaniments of distress, harrowed up her soul, and she resolved to try Col. Lightfoot once more, though mortifying, to tell him the cause of her importunities. She gained his presence the next morning, and found upon his brow that peculiar look, which indicated unwillingness to hear any more from her upon the subject of her errand. But she entreated him to listen to her motives, and unfolded to him her situation; that although her son was but a boy, he was her gallant protector and defender."

This confession changed matters completely, and Lightfoot instantly said, "Go madam, and take your son. His great worth had caused me to desire much his management of my business; but your need is entitled to the preference; and those rare qualities and powers, which he possesses, and which had gained him my confidence and esteem, will ensure your protection."

3. Stephen B. Weeks, "Benjamin Hawkins," in Ashe, *Biographical History of North Carolina,* 5: 144–53.

4. John D. Hawkins, "Colonel Philemon Hawkins, Sr."

5. Details are given in the introduction to the Hawkins Family Papers. The eldest six Hawkins children, born in North Carolina, were Sallie (1837–60), Willis (1839–63), Virginia (1840–1930), James Jr. (1842–58), John Davis II (1844–65), and little Ariella, called Ella (1846–51). About 1846, when little Ariella was a baby in arms, James B. Hawkins and Ariella and their six North Carolina–born children came by Mississippi riverboat to Texas, where their two Texas children, Frank (1847–1901) and Edgar (Charles Edgar, 1852–87), were born.

6. Names and activities of the Alston family are given in the introduction to the Archibald D. Alston Papers. Ariella's siblings were Charles Julian Poydroas Alston (b. 1818); Leonidas (1823–49); Missouri (b. 1824), and Edgar (1827–48). Her year of birth is uncertain. Her obituary in the *Velasco World,* dated March 15, 1902, gives her age as eighty-five, making her birth year 1817. However in the Matagorda Census of 1850, her age is put at twenty-nine, making her year of birth 1821. The 1821 birth year (January 26, 1821) is also given in Graves, *The Alstons and the Allstons of North and South Carolina,* 131. If the later birth year is correct and if she and James B. Hawkins married on January 16, 1836, as records indicate, her age at her marriage was not quite sixteen.

7. James B. Hawkins to Dr. William J. Hawkins, January 15, 1847, Archibald D. Alston Papers. On navigability of Caney Creek, see Edgar M. Alston to Mrs. Sarah Alston, written from Galveston, December 22, 1845, Archibald D. Alston Papers.

Chapter 2: Letters Written en Route

1. Ariella Alston to Mrs. Sarah Alston, November 20, 1846, Archibald D. Alston Papers.

2. Ibid.

3. Ann Read Hawkins to Mrs. John Davis Hawkins, October 14, 1848, Hawkins Family Papers.

4. Ibid.

5. Frank Hawkins to John Davis Hawkins, November 1, 1848. Hawkins Family Papers.

6. Frank Hawkins to John Davis Hawkins, November 14, 1848, Hawkins Family Papers.

7. Frank Hawkins to John Davis Hawkins, November 29, 1848, Hawkins Family Papers.

8. Ibid.

Chapter 3: Starting the Caney Sugar Plantation

1. Partnership Agreement of James B. Hawkins and John Davis Hawkins, Jr., October 17, 1846, Hawkins Ranch, File of Historical Data.
2. James B. Hawkins to John D. Hawkins, January 15, 1847, Hawkins Family Papers.
3. James B. Hawkins to Dr. William J. Hawkins, January 16, 1847, Hawkins Ranch, File of Historical Data.
4. Ariella Alston Hawkins to Sarah Alston, February 27, 1847, Archibald D. Alston Papers.
5. James B. Hawkins to Dr. William J. Hawkins, January 15, 1847, Hawkins Ranch, File of Historical Data.
6. Ibid.
7. Ariella Alston Hawkins to Sarah Alston, December 26, 1847, Archibald D. Alston Papers. James Hawkins spelled the neighbor's name Steward.
8. Ibid.
9. James B. Hawkins to Dr. William J. Hawkins, January 16, 1847, Hawkins Ranch, File of Historical Data.
10. James B. Hawkins to John D. Hawkins, December 25, 1848, Hawkins Ranch, File of Historical Data.
11. James B. Hawkins to Major Archibald D. Alston, January 12, 1849, Archibald D. Alston Papers.
12. Ariella Alston Hawkins to Sarah Alston, November 9, 1849, Archibald D. Alston Papers.
13. Lamar, *Charlie Siringo's West,* 58.

Chapter 4: Ariella and Plantation Family Life

1. Ariella Alston Hawkins to Sarah Alston, February 27, 1847, Archibald D. Alston Papers. The Jones mentioned is possibly John H. Jones, whose plantation on Caney is cited in Ingram, *Canebrake Settlements.* Salem Female Academy was founded in 1802 by a Moravian settlement, itself founded in 1765.
2. Ariella Alston Hawkins to Sarah Alston, February 27, 1847, Archibald D. Alston Papers.
3. Ariella Alston Hawkins to Sarah Alston, November 9, 1849, Archibald D. Alston Papers.
4. Ibid.
5. Ariella Alston Hawkins to Sarah Alston, December 11, 1850, Archibald D. Alston Papers.
6. Sallie Hawkins to Sarah Alston, May 6, 1849, Archibald D. Alston Papers.
7. Ariella Alston Hawkins to Sarah Alston, April 11, 1847, Archibald D. Alston Papers.
8. Ariella Alston Hawkins to Sarah Alston, February 27, 1847, Archibald D. Alston Papers.
9. Ibid.
10. Ariella Alston Hawkins to Sarah Alston, April 11, 1847, Archibald D. Alston Papers.
11. Ariella Alston Hawkins to Sarah Alston, November 9, 1849, Archibald D. Alston Papers.
12. Sallie Hawkins to Sarah Alston, May 6, 1849, Archibald D. Alston Papers.
13. James B. Hawkins to Sarah Alston, April 28, 1850, Archibald D. Alston Papers.
14. Ariella Alston Hawkins to Sarah Alston, December 11, 1850, Archibald D. Alston Papers.

Chapter 5: The Case of Edgar and Ways of Thought in Slavery Times

1. Campbell, *An Empire for Slavery,* 67.
2. Ibid., 78, and appendix 2 on slave populations, 264–65.
3. Edgar M. Alston to Sarah Alston, December 22, 1845, Archibald D. Alston Papers.
4. Ibid.
5. Ibid.
6. Ariella Alston Hawkins to Sarah Alston, June 26, 1846, Archibald D. Alston Papers.
7. Memucan Hunt to Sarah Alston, November 10, 1846, Archibald D. Alston Papers.
8. Ariella Alston Hawkins to Sarah Alston, February 27, 1847, Archibald D. Alston Papers.
9. Edgar M. Alston to Sarah Alston, August 25, 1847, Archibald D. Alston Papers.
10. James Boyd Hawkins to Charles Alston, n.d., Archibald D. Alston Papers.
11. Ibid.
12. Campbell, *Gone to Texas,* 269.
13. Ariella Alston Hawkins to Sarah Alston, December 26, 1847, Archibald D. Alston Papers.
14. James Boyd Hawkins to Major Archibald Alston, June 12, 1849, Archibald D. Alston Papers.
15. James B. Hawkins to Dr. William Hawkins, January 15, 1847, Hawkins Family Papers.
16. James B. Hawkins to John D. Hawkins, January 15, 1847, Hawkins Family Papers. Business transactions seem to be put in family letters as reminders of previously agreed-on obligations, understood by both correspondents. Letters alone do not give us the necessary detail to calculate the obligation with any accuracy. Not infrequently such business transactions mentioned in letters request that a third party, usually a family member, be paid some amount out of the proceeds of a sale. This informal linkage of credit and debit indicates the scarcity of banking services.
17. John D. Hawkins Jr. to Charles Alston, October 31, 1858, Archibald D. Alston Papers.
18. Hawkins Ranch, File of Historical Data.
19. Ariella Alston Hawkins to Missouri Alston, n.d., Archibald D. Alston Papers.
20. Ann Read Hawkins to Mrs. John Davis Hawkins, October 14, 1848, Hawkins Family Papers.

Chapter 6: Building the Ranch House (Lake House), 1854

1. James B. Hawkins to Sarah Alston, January 12, 1854, Archibald D. Alston Papers. In family correspondence in this collection there are four letters referring to the building of the Hawkins Ranch House begun in 1854 and then called the Lake House. The first, quoted in the chapter, is the January 12 J. B. Hawkins letter to Mrs. Alston already cited. The second (February 10, 1854) is a letter from Mrs. Alston to her young grandson Willis Alston, son of Ariella's sister, Missouri Alston: "Aunt Ella is quite well, [She has been ill], walks and rides about everywhere. Uncle Jim has commenced building a fine House on Lake Austin. No sickness on the plantation."

The third (March 22, 1854) is J. B. Hawkins's letter to Major Archibald D. Alston, also quoted in the chapter. The framing of the house is up, and J. B. Hawkins describes the house, its rooms, the number of stories, and details of fireplaces and closets—all of which match the details of the house today. And the fourth (November 24, 1854) is J. B.

Hawkins's description of the hurricane of 1854 and the damage done to the Hawkins Ranch House, the town of Matagorda, and the chimney of his sugar house on Caney. The kitchen of the ranch house was "blown flat." The house itself was "partly raised [razed]," and a man on the place was seriously injured.

2. From a sample, the flooring was identified as ash by the Center for Wood Anatomy Research, United States Department of Agriculture, Forest Service, Madison, Wisconsin, July 16, 2008.

3. Thomas Decrow had a two-story dwelling built especially to withstand Gulf Coast storms. It was "fastened together" by an expert ship's carpenter; see "Antebellum Social Conditions," ch. 9 in Matagorda County Historical Commission, *Historic Matagorda County,* 162.

4. Wayne Bell, interview with author, Austin, Texas, 2007.

5. James B. Hawkins to Major Archibald D. Alston, March 22, 1854, Archibald D. Alston Papers.

6. The measurements of the plantation house and the Hawkins Ranch House were supplied in 2007 by Bill Isaacson, business manager of the Hawkins Ranch Ltd. The plantation house, though now in ruins, has a visible footprint that allowed measurement.

On February 2, 2006, Mary Belle Ingram of Bay City, Texas, the local archivist and Matagorda County historian, confirmed by her research of county records that the plantation house "bought of Quick," which had to be "patched up" for the Hawkins family's use when they arrived in 1846, was the only house occupied at Caney by the Hawkins plantation family. Its location was (and the ruin of the house is) in the Kingston and Powell league. Mrs. Jacob Quick, from whom J. B. Hawkins bought the house, was the daughter of William Kingston, who received this sitio of land in 1827. He conveyed it to his daughter, Catherine Quick (Mrs. Jacob Quick), on January 18, 1842. It included two hundred acres of land and was conveyed to J. B. Hawkins by her on March 24, 1846. This plantation house and land became part of the estate of J. B.'s youngest son, Edgar Hawkins, and his heirs; consequently following the death of J. B. Hawkins and Ariella, this house and surrounding acres were not within the Hawkins Ranch land conveyed to Frank Hawkins and subsequently to his five children.

The orange trees in the Hawkins Ranch House yard were planted by Ariella. They were a gift from her husband's brother, Dr. Alexander Hawkins, who in April 1858 married Martha Bailey, whose family owned a citrus orchard in Florida. Dr. Hawkins and his wife moved to Leon County, Florida, where he became a citrus grower himself, and in 1884 began the cultivation of a variety of sweet oranges and grapefruit. By 1894 he abandoned citrus growing because of a severe frost that devastated the trees. He then returned to North Carolina and lived at Raleigh. Samuel A. Ashe, "Alexander Boyd Hawkins," in Ashe, *Biographical History of North Carolina,* 5: 164–68. The year the orange trees were planted in the Hawkins Ranch House yard is not known, but the root stock of the trees has been alive since Ariella's first planting, and the trees continue to bear fruit.

7. "Antebellum Social Conditions," ch. 9 in Matagorda County Historical Commission, *Historic Matagorda County,* section on hurricanes, 161–62.

8. James B. Hawkins to Sarah Alston, November 24, 1854, Archibald D. Alston Papers.

9. See "John Duncan," in Matagorda County Historical Commission, *Historic Matagorda County,* 64.

10. Sallie Hawkins to Jane, her cousin in North Carolina, August 24, 1857, Archibald D. Alston Papers.

Chapter 7: Effects of Civil War and Emancipation

1. Campbell, *Gone to Texas,* 260.
2. Hawkins Ranch, File of Historical Data.

3. Ibid.; Fehrenbach, *Lone Star,* 359.

4. Lamar, *Charlie Siringo's West,* 25, and see also ch. 2, "The Civil War Comes to Matagorda Bay, 1850–1867," 17–32.

5. Matagorda County Historical Commission, *Historic Matagorda County,* 165, and ch. 10 on the Civil War, 163–74.

6. Ralph J. Smith, quoted in Gallaway, *Texas, the Dark Corner of the Confederacy,* 189.

7. Francis Richard Lubbock, quoted in Gallaway, *Texas, the Dark Corner of the Confederacy,* 176.

8. Hawkins Ranch, File of Historical Data; on sending more laborers see John G. Forister and Hershel Horton, "Confederate Defenses at the Mouth of the Caney," *Bay City Tribune,* November 29, 2009.

9. Campbell, *Gone to Texas,* 268–69.

10. Fehrenbach, *Lone Star,* 396; the freed slave's remark is quoted in Campbell, *Gone to Texas,* 269.

11. Campbell, *Gone to Texas,* 347. See also "Prison System," in *Handbook of Texas Online,* http://www.tshaonline/articles PPjjp3.html, accessed September 2, 2010.

12. Barr, *Black Texans,* 91.

Chapter 8: Frank Hawkins and the Development of Cattle Ranching

1. William A. Kirkland to J. C. Lewis, September 11, 1968, Hawkins Ranch, File of Historical Data. William Kirkland forwarded a notation of his aunt, Rosa Noyes (Mrs. L. T. Noyes), who was born in 1850 in Mississippi and died in 1948. She remembered that her brother Jesse Kirkland (William Kirkland's father's half brother) and Frank Hawkins had gone to school together in Germany. The school she named was Carlshaven near Cassel, Germany. Rosa Noyes lived on Caney Creek from 1861 to 1868, and she remembered that Ariella had planted the oak trees in the front yard of the Hawkins Ranch House. The date of the planting is not stated.

2. Hawkins Ranch, File of Historical Data.

3. Virginia Hawkins to Charlie Alston, July 14, 1869, Archibald D. Alston Papers.

4. Dance Card of Allesley, December 1869; letter of M. E. Willis to Mr. Hawkins from Badminton House, n.d.; clothing bills of 1870. The clothing Frank Hawkins bought included gloves from Geo. H. Haywood, Hosier, Glover, Shirt & Collar Maker etc. at #3 Earl Street, Coventry. The bill was dated May 21, 1870. Frank also was charged on April 6 and June 3, 1870, for coating and tweed from David Spencer & Company. Hawkins Ranch, File of Historical Data.

5. Receipts for their order of furniture in Hawkins Ranch, File of Historical Data.

6. Jones, *Texas Roots,* ch. 8, "Hunting and Stock Raising," 171–94.

7. Matagorda County Historical Commission, *Historic Matagorda County,* 184, and see ch. 11, "Economic Development of the County Prior to 1890, the Cattle Industry," 174–91.

8. Ibid. For cattle drives to market, see Siringo, *A Texas Cowboy,* and Lamar, *Charlie Siringo's West.*

9. Campbell, *Gone to Texas,* 214: "The constitution of 1845 prohibited banking in Texas and required every corporation to obtain a charter from the legislature, thereby creating a serious deterrent to economic expansion in the new state." For banking history in Bay City, Texas, see Matagorda County Historical Commission, *Historic Matagorda County,* 487–89.

Chapter 9: Ariella's Fight for Her Rights

1. A copy of James B. Hawkins's will and other deeds to Frank Hawkins and Virginia Hawkins Brodie are in Hawkins Ranch, File of Historical Data.

2. R. C. Duff to Mrs. J. B. Hawkins, October 17, 1899, Hawkins Ranch, File of Historical Data. The story of Ariella's contemplated lawsuit against her son Frank Hawkins to reclaim her property rights is detailed in a collection of thirteen letters she exchanged with her attorneys, F. J. and R. C. Duff, in 1899 and 1900. Only the attorneys' letters are dated. Ariella's undated letters are in pencil, written on scraps of paper, and appear to be rough drafts she may have sent to her attorneys later in final form. In some exchanges with her lawyers, her words are known only because her lawyers quote them back to her in their replies. This collection of letters, labeled "Ariella Alston Hawkins and the Duffs: Their Correspondence, 1899–1900," is in the Hawkins Ranch, File of Historical Data, as are all letters and documents referenced in this chapter; the thirteen-letter collection is hereafter cited as Ariella and the Duffs.

3. Memorandum of agreement between Ariella Hawkins and the attorneys Duff, copy in Ariella and the Duffs.

4. The demeanor of Ariella and Frank Hawkins is described by R. C. Duff's letter to Ella A. Hawkins, December 11, 1899, and by Mr. Austin, quoted in that letter, in Ariella and the Duffs.

5. Ariella Hawkins letter, n.d., in Ariella and the Duffs.

6. F. J. and R. C. Duff to Mrs. James B. Hawkins, November 25, 1899, in Ariella and the Duffs.

7. F. J. and R. C. Duff to Mrs. James B. Hawkins, February 4, 1900, in Ariella and the Duffs.

8. Ibid.

9. R. C. Duff to Mrs. Ella A. Hawkins, December 5, 1899, in Ariella and the Duffs.

10. R.C. Duff to Mrs. Ella A. Hawkins, December 11, 1899; Ariella's deed in her own handwriting is in Hawkins Ranch, File of Historical Data.

11. The letter of December 11, 1899, in which the Duffs try very hard but in vain to justify their professionalism to Ariella, is typed single-spaced on fourteen half sheets. Ariella and the Duffs.

12. Miss Tenie Holmes, personal communication with author.

Chapter 10: A Birth, a Death, and the Move to Town, 1896

1. Rowland Rugeley, personal communication with author. That Elmore died in the west bedroom is a supposition, but my brother Frank had information that this was the room occupied by Frank and Elmore Hawkins. The south bedroom (Meta's) was the one occupied by James B. and Ariella. Whether Dr. Rugeley attended his daughter in her delivery or she was cared for by a midwife is not known; we can only guess that had he attended her, he would probably still have been present when her postpartum condition worsened.

2. Elmore Rugeley Hawkins was buried in the Rugeley family cemetery on Caney Creek near Van Vleck, the place of the Rugeley's plantation home. The burial was on a Sunday afternoon in April 1896 and according to the newspaper notice was "the largest funeral procession ever seen in the county." At the death of their mother on April 3, 1896, Harry was eight, Meta six, Janie four, Lizzie two, and Elmore (called Sister) newborn.

3. These descriptions derive from photographs from the album of James B. Brodie, 1899. Because of the faded condition of these album photographs, they are not included among the illustrations in this book.

CHAPTER 11: SCHOOLING AND A HOUSE OF THEIR OWN, 1913

1. News clipping, December 1900, Larry Phillips Young Collection, Matagorda County History Museum, furnished in 1998 by Mary Belle Ingram. One of Mrs. Holmes's stories quoted in the clipping related to a mysterious experience she had of earning a little extra money as a seamstress: "Under a certain bush she would find cloth with measurements and instructions for making uniforms, and orders to place the completed garments under the same bush on a certain night. This practice kept up for some time, and she never saw or knew who furnished the cloth or got the uniforms." Mrs. Holmes's patron was a member of the Ku Klux Klan, which had gained adherents in the county during Reconstruction.

2. *Daily Tribune,* Bay City, Texas, December 1900, *Tribune* file, Matagorda County Museum.

3. Details are from Meta's scrapbook and Kidd-Key Album of 1909 in the author's possession.

4. Whitis School Archive, Austin History Center, Austin Public Library, Austin, Texas.

5. Ibid.

6. Copy of the Trustees Petition to the District Court, Matagorda County, 1917, Hawkins Ranch, File of Historical Data.

7. Mrs. Effie Dickey, quoted in Stieghorst, *Bay City and Matagorda County,* 130.

8. Hawkins Ranch, File of Historical Data.

CHAPTER 12: YOUNG LADY RANCHERS IN CHARGE, 1917

1. Hawkins Ranch, File of Historical Data: "Statement of Cash Receipts and Disbursements for the Month of May 1915. F.A. Bates, the Bookkeeper. His fee $10 per month. On the sale of cattle which took place May 4, 1915, the amount received was $7,495.00." A note at the bottom of the statement reads:

> We branded all calves large enough this spring, beginning May 17th. Total branded 654 against 764 last spring, but last year we began June 19th, or one month later, which I think accounts for this shortage. This will no doubt be offset by increased number of calves in Nov. branding. It is impossible to tell exactly what effect Charbon last year may have had but no doubt it was considerable, on this year calf crop.
>
> I sold what yearlings we had with a few two year old steers at $25 for yearlings and $35 for twos. This being the best price ever received for this class. You will note increase of $10.00 for each year in these cattle after one year old, but from birth to first year $25.00 therefore I think best to hold calves to yearlings rather than yearlings to twos. We have previously shipped a great many calves during summer, but I now advise holding calves, as the price of calves seems to be getting lower, and older cattle advancing.
>
> All heifer yearlings with mothers have been separated and weaned, and placed in pastures by themselves, we had 500 of these of full year old in pen at one time, which experienced stockmen said were the prettiest lot of cattle they had ever seen in this county. These were almost entirely ¼ Brahma. The value of the Brahma is being so well recognized that the Brazilian Government has recently paid $12000.00 each for pure breds imported from India, for the purpose of improving native breeds.

2. Copy of the Trustees Petition to the District Court of Matagorda County, 1917, Hawkins Ranch, File of Historical Data.

Chapter 13: Courtship and Marriage

1. *Daily Tribune,* Bay City, Texas, March 1915, clipping in family album in author's collection.

2. *Daily Tribune,* Bay City, Texas, October 1917, clipping in family album in author's collection.

3. J. C. Lewis to Meta Hawkins, March 1917, family album in author's collection.

4. Personal communication with author, 1940s.

5. Berkeley Holman, quoting his father's advice to Esker McDonald, personal communication with author, ca. 1980.

6 . Elmore Hawkins McDonald, with comic exaggeration, personal communication with author, ca. 1960. The *Daily Tribune,* on September 3, 1927, published the following notice of the marriage: "Culminating a romance extending over a period of eight years, Miss Elmore Hawkins, yesterday, became the bride of Mr. Esker L. McDonald. The ceremony was performed at 7:15 o'clock p.m. at the second Baptist Church in Houston, Rev. M. M. Wolfe officiating. The Bride wore a becoming grey ensemble with American Beauty hat and accessories in harmony, and carried an arm cluster of bride's roses. They were attended by Mr. and Mrs. M. J. Murphy. . . . Following the ceremony the happy couple was entertained with a dinner at the Rice Hotel, given by Mr. and Mrs. Murphy, after which they left on the Sunshine Special for Chicago and other points of interest."

Chapter 14: Lizzie

1. Farm Road 521 is a state road properly called Farm to Market Road 521, in roadside signage F.R. 521. It was authorized in 1945, and when this two-lane paved highway was constructed, it cut across the Hawkins Ranch from west to east, significantly separating the Sheppard Mott pasture from the main Hawkins Ranch—an unwelcome development, the ranch owners thought. F.R. 521 runs for about ninety-five miles serving Matagorda, Brazoria, Fort Bend, and Harris counties.

2. Throughout the chapters in part II, quoted dialogue drawn only from my memory is not usually formally sourced; but each quotation is faithful to the personally communicated voice of the speaker.

3. After Lizzie's death my mother and her sisters had the sad duty of entering Lizzie's house for the first time in decades because she had refused to allow it. Now, feeling intrusive, they had to enter uninvited, to learn what bills needed paying. My mother described to me, shortly after Lizzie's death in 1957, how the crammed pigeonholes of her desk told the story of her mounting unpaid bills.

Chapter 15: The Conversations in the Family, 1935

1. The name regularly used for the house by the Hawkinses was the Currie house or the Currie place, because the Hawkins Ranch leased land the Curries owned. Properly, the name should be the Cavanah-Currie house, because the Cavanah family, original Stephen F. Austin settlers, were its first residents and owned the house at the time of the massacre.

2. The massacre is described in Matagorda County Historical Commission, *Historic Matagorda County,* 42. The victims indicated are the Cavanah and Flowers families. The article says the house was occupied at the time by the Cavanah family and that members of the Flowers family were visiting when the massacre took place. The sources for this article are the oral reports in 1927 of Jesse Matthews and my aunt, Janie Hawkins, to John C.

Marr for Marr's unpublished 1928 thesis, *The History of Matagorda County, Texas.* The Hawkinses of my memory always identified the house and location as "the Currie place" because the Currie family occupied it within their childhood and owned the surrounding acres for many decades, until 2002.

The handed-down stories of the massacre at the Cavanah-Currie house have some of the qualities of an eyewitness account and many of story-teller's embellishment. Texas settlers complained to Stephen F. Austin about attacks by the Karankawas, and he pursued them with a military force. Some of these complaints may have included details of what happened at the Cavanah-Currie house. Early characterizations of the Karankawa tribe (or group of tribes) render them as fierce, cannibalistic savages. Stephen F. Austin accepted this characterization, although his approach was generally to use friendly gestures toward them to avoid provocation. Cantrell, *Stephen F. Austin,* 96–97, 136–39; Barker, *Life of Stephen F. Austin,* 94.

An example of regarding the Karankawas as savage is John Wesley Wilbarger's *Indian Depredations in Texas,* published 1889, in which the author says his motive in writing is "to preserve in history the story of massacres and conflicts with Indians" as a memorial to his brother, who was scalped. He claims to base his accounts of massacres on eyewitness accounts but does not say whose accounts they are. Wilbarger's story of the Karankawa massacre at the Cavanah-Currie house on the Hawkins Ranch (*Indian Depredations in Texas,* 209–10) was repeated by Ed Kilman in his popular 1959 book *Cannibal Coast.* To the best of my knowledge, most retold stories of the Cavanah-Currie massacre derive from Wilbarger and Kilman. Closely following Wilbarger, Kilman (*Cannibal Coast,* 242) describes the massacre this way:

> In the winter of 1830 about seventy Karankawas swooped down upon the lonely cabin of Charles Cavanaugh on Caney Creek. Cavanaugh was away and his wife and children were entertaining Mrs. Elisha Flowers and her daughters, their neighbors. The yelling savages stormed into the peaceful house and killed Mrs. Cavanaugh and three of her daughters. They wounded a fourth daughter and threw her into a brush pile near the house, leaving her there to die. Mrs. Flowers, trying to escape with her child, was run down and slain. The Indians tomahawked her daughter and tossed her body into the brush pile with the Cavanaugh girl. . . . When Cavanaugh returned home, he was crazed by the ghastly scene before him. For a while he gibbered like a madman. Then his mind clearing, he swore by the memory of his loved ones whose mutilated bodies lay cold around him that he would dedicate the rest of his life, if need be to the extermination of the Karankawas.

More recent historical and anthropological studies take a revisionist view that questions the essential cannibalism of the Karankawas: see Campbell, "Karankawa Indians"; Ricklis, *The Karakawa Indians of Texas;* and La Vere, *The Texas Indians.* La Vere treats the Karankawas in the chapter 3 section "The Arrival of Strangers" (59–64) and, like Riklis, finds that "little direct evidence bolsters the claim of Karankawa cannibalism" (62).

Chapter 16: Janie and Harry

1. The Hawkins family's artist friend Georgia Mason Huston is referred to several times in this book. During her lifetime she was a steady participant at Thanksgiving, Christmas, or other gatherings. On first acquaintance she lived in Houston with her husband, Joe Mason, and often visited Bay City and took part in country excursions or bird shoots. She painted several scenes of the Hawkins Ranch, including the plantation, the Ranch House, and Lizzie's headquarters. Divorced, she move to Bay City and married P. G. Huston, family friend and local pharmacist.

Chapter 17: Sister and Esker

1. The MacDonald-Nelson 1935 film *Naughty Marietta* would have provided Sister and Esker with Victor Herbert's "Ah Sweet Mystery of Life."

2. Assuming that the Hawkins children were taken to their grandparents the year of their mother's death, their respective ages would have been Harry, eight; Meta, five; Janie, four; Lizzie, two; and Sister, newborn.

Chapter 18: Meta and Jim

1. Meta's bridge foursome is referred to numerous times in these pages. As a child, coming home from school, I found the conversation of the bridge players gave me news of the adult world. The foursome had no fixed membership. At one time or another it consisted of any four available players among close friends: Daughty Rugeley, Irby Stinnett, Ruby Sanborn, Jerry Winston, Stella Collins, Lurline Wadsworth, Thelma Harrington, and Leola Moore.

2. Rose and Douglas Chapman as well as other long-standing African American cooks, helpers, and ranch hands are featured in chapter 25, "Good People on the Place."

Chapter 19: Rowland and Daughty

1. C. K. Norcross and his work on the Hawkins Ranch House are more fully described in chapter 21. The Ranch House is pictured in the book's frontispiece as it looked in 2008, seventy-three years after the Norcross renovation, but the essential features with the arch of oak trees are the same. The house was built, as has been said, in 1854. How long after that date the original oak trees were planted is not known. From time to time some oaks have been replaced.

Chapter 20: The Lady Visitor and the Decision

1. Vera Prasilove (1899–1966), also known as Vera Prasilove Scott, began her professional training in photography in Prague at age eighteen and continued academic studies in Munich, where she met her future husband Arthur F. Scott while he was doing postdoctoral research in chemistry. Married in 1925, she went with her husband to Houston, where he joined the faculty of Rice University (then Rice Institute), and she established a photographic studio specializing in portraiture. During her Houston years (1926–37) she did portraits of many well-known Houston families and Rice faculty. Many of her works are in the permanent collections of the Portland Museum of Art and the Museum of Czech Literature in Prague. Rice University holds a collection of her portraiture (1926–37) in the Amanda Focke Woodson Research Center of the Fondren Library. Websites: http://library.rice.edu/collections/WRC/finding-aids/manuscripts/0497, http://www.lib.utexas.edu/taro/ricewrc/00176.xml.

2. After Farm to Market Road 521 was built in 1945, it provided easier access to the Hawkins Ranch; the dirt road through the Sheppard Mott from the Liveoak road was no longer the customary route.

Chapter 21: The Ranch House and Mr. Norcross

1. Mining operations and employees were transferred to a new location, the company town New Gulf in Wharton County, Texas. See "Sulphur Mining in Matagorda County, 1919–1932," in Matagorda County Historical Commission, *Historic Matagorda County,* 339–42.

2. Delco made battery ignition systems for automobiles and developed the Delco-Light system, an internal combustion generator to provide electricity for homes not served by the power grid; see http://delcolight.com/20.html. The Hawkins Ranch system was adapted for the windmill to power it.

3. As indicated in a chapter 6 note, floor material was confirmed as ash. Much of the material in the Hawkins Ranch House is oak, according to Logic Tobola, architect for the 1999 renovation of the house. My mother, Meta Hawkins Lewis, and Helen Cates Neary of Bay City, both knowledgeable about the construction of nineteenth-century houses in Matagorda County, said that in addition to local woods, some lumber was brought by boat from New Orleans. The skillfully turned wood of the banister in the Hawkins Ranch House may have been such an import.

4. This feature can give the impression that the stair is a continuous serpentine, but it is in fact two straight stairways, with a continuing balustrade and handrail. Robert James Coote, architect, personal communication with author, September 26, 2013.

Chapter 23: The Alley Way

1. http://www.pinetreeweb.com/1937-nj1–05.htm. The jamboree was held from June 30 to July 9, 1937, drawing some twenty-five thousand scouts, mainly from the United States but many from foreign countries; they camped in tents along the Potomac within sight of the Washington Monument.

Chapter 24: Miss Tenie

1. E. Hawes Sr., quoted in Guthrie, *Forgotten Ports,* 189–97: "To D. Lewis of Indianola, or any merchant if he is not there. Please send me five barrels of flour. We have nothing to eat on the island except meat. All people on the island saved except Betty Mead (colored) and her two children, all houses except the life saving station at the upper end of the island washed away and destroyed, and people left with only the clothes they had on their back. Nearly all the cattle and sheep were drowned, and the balance will die for want of water." The importance of Saluria in the coastal defense of Texas during the Civil War is from Fitzhugh, "Saluria."

Chapter 25: Good People on the Place

1. Lewis, "The Range Cattle Industry in the Southwest."

2. *San Antonio Express-News,* February 2, 1986.

3. Ibid.

4. Harkell Bend, the southernmost area of the Hawkins Ranch, lies east of Cottonwood Creek and rises to a point at the south border of the Pamela Pickett league. It includes land within the "boot" configuration of Liveoak Creek (see ranch map).

5. *San Antonio Express-News,* February 2, 1986.

Chapter 26: Frank Hawkins Lewis, Cattleman

1. Lewis, "Notes for a Class Reunion," n.d., Hawkins Ranch, File of Historical Data.

2. Major T. C. Eastman to Frank H. Lewis, September 13, 1942; letter of commendation for voluntary service on maneuvers, Hawkins Ranch, File of Historical Data.

3. Frank H. Lewis to Mr. and Mrs. J. C. Lewis, February 11, 1945, Hawkins Ranch, File of Historical Data.

Chapter 27: The Future of the Sense of Place

1. The already announced course was "Mysticism and Existentialism" in the Department of Religion at Rice University. The department chair was aware that my dissertation treated philosopher-theologians relevant to the two idea groups.

2. In 1937, independently of J. C. Lewis, Esker McDonald bought the River Ranch as a long-range investment and place of recreation, where he raised a small herd of Charolais cattle. It is a pretty, tree-shaded place along the east bank of the Colorado River just north of Bay City. About 1,383 acres of the original purchase eventually came into the holdings of the Hawkins Ranch partnership.

The Buckeye Ranch was a joint purchase in 1944 by Lewis and McDonald. They farmed rice there and in that way, over time, made the land pay for itself. The property, located west of Bay City and west of the Colorado River, belonged to the Stoddard and Kreger families, who had the idea, never fulfilled, of creating a town. The place was named for their native "buckeye state," Ohio. The land coming into the holding of the Hawkins Ranch partnership amounted to about 5,000 acres. Other properties the two men acquired were the Peyton Creek place, a farm of 1,599 acres adjacent to the Sheppard Mott; and the Slough, a farm of 2,556 acres, on Tres Palacios Bay near the town of Palacios. It was sold in 2001.

3. The ten partners were Frank H. Lewis and offspring, Frank H. Lewis Jr., Janet Lewis Peden, Meta Lewis Hausser, and James N. Lewis; and Margaret Lewis Furse and offspring, Jane H. Furse, Austen H. Furse III, John L. Furse, and Mary E. Furse.

4. By the time of writing, the interest energy companies once expressed in locating wind farms in the vicinity had subsided.

Appendix: Sketches and Letters of the Antebellum Children

1. Photographs and details of the history of Patapsco Female Institute are at www.ellicottcity.net/tourism/attractions/patapsco and www.patapscofemaleinstitute.org. "Almira Hart Lincoln Phelps" is the subject of episode 2032 of the radio series Engines of Our Ingenuity, by John H. Lienhard. At sixty-six, Phelps was the second woman to become a member of the American Association for the Advancement of Science. She is profiled in Ogilvie, *Women in Science,* 147–48. On the preparation of women to assume unfamiliar responsibilities during the Civil War, see Faust, *Mothers of Invention.*

2. John D. Hawkins to Charles Alston, written from Henderson Depot, N.C., October 31, 1858, Archibald D. Alston Papers.

3. Virginia Hawkins to Sarah Alston, November 18, 1860, Archibald D. Alston Papers.

4. Ibid.

5. Leah Hawkins to Sarah Alston, n.d. [1863, 1864?], Archibald D. Alston Papers.

6. E. G. Brodie to a friend, May 16, 1861, Archibald D. Alston Papers.

Bibliography

Primary Sources

Archibald D. Alston Papers, #16, Southern Historical Collection, Wilson Library, University of North Carolina at Chapel Hill.

Hawkins Family Papers, #322, Southern Historical Collection, Wilson Library, University of North Carolina at Chapel Hill.

Hawkins Ranch, File of Historical Data, Hawkins Ranch Ltd., 2020 Avenue H, Bay City, Texas 77414.

Matagorda County Museum, Courthouse Square, Bay City, Texas.

Whitis School Archive, Austin History Center, Austin Public Library, Austin, Texas.

Secondary Sources

Alexander, Drury Blakeley. *Texas Homes of the Nineteenth Century.* Austin: University of Texas Press, 1966.

Allen, Arda Talbot. *Twenty-One Sons for Texas* (Rugeley/Hawkins genealogy). San Antonio: Naylor, 1959.

Ashe, Samuel A'Court. *Biographical History of North Carolina,* vol. 5: *From Colonial Times to the Present.* Greensboro: Charles L. Van Noppen, 1906.

Baker, T. Lindsay, and Julie P. Baker, eds. *Till Freedom Cried Out: Memories of Texas Slave Life.* College Station: Texas A&M University Press, 1997.

Ball, Edward. *Slaves in the Family.* New York: Ballantine, 1998.

Barker, Eugene C. "The Influence of Slavery in the Colonization of Texas." *Southwestern Historical Quarterly* 28 (July 1924): 1–33.

———. *The Life of Stephen F. Austin.* 1927; reprint, Texas History Paperback, Austin: University of Texas Press, 1969.

Barr, Alwyn. *Black Texans: A History of African Americans in Texas 1528–1995,* 2nd ed. Norman: University of Oklahoma Press, 1996.

Collodi, Carlo. *The Adventures of Pinocchio.* 1883 (in Italian); English translation by Mary Alice Murray, 1892, London: Everyman's Library, 1911.

Campbell, Randolph B. *Gone to Texas: A History of the Lone Star State.* New York: Oxford University Press, 2003.

Campbell, Randolph B. *An Empire for Slavery: The Peculiar Institution in Texas,* 1821–1865. Baton Rouge: Louisiana State University Press, 1989.

Campbell, Randolph B., and Richard Lowe. "Some Economic Aspects of Antebellum Texas Agriculture." *Southwestern Historical Quarterly* 82 (April 1979): 351–78.

———. *Wealth and Power in Antebellum Texas.* College Station: Texas A&M University Press, 1977.

Campbell, T. N. "Karankawa Indians." In *The Handbook of Texas,* vol. 3 edited by Eldon Stephen Branda, 464–65. Austin: Texas State Historical Association. *The Handbook of Texas Online,* http://www.tshaonline.org.

Cantrell, Gregg. "The Partnership of Stephen F. Austin and Joseph H. Hawkins." *Southwestern Historical Quarterly* 99, no. 1 (July 1995): 1–24.

———. *Stephen F. Austin: Impresario of Texas.* New Haven, Conn.: Yale University Press, 1999.

Clarke, Erskine. *Dwelling Place: A Plantation Epic.* New Haven: Yale University Press, 2005.

Culbertson, Margaret. *Texas Houses Built by the Book: The Use of Published Designs, 1850–1925.* College Station: Texas A&M University Press, 1999.

Curlee, Abigail. "The History of a Texas Slave Plantation 1831–1863." *Southwestern Historical Quarterly* 26 (October 1922): 79–127.

———. "A Study of Texas Slave Plantations, 1822–1865." Ph.D. diss., University of Texas, 1932.

Faust, Drew Gilpin. *Mothers of Invention: Women of the Slaveholding South in the American Civil War.* Fred Morrison Series in Southern Studies. Chapel Hill: University of North Carolina Press, 1996.

Fehrenbach, T. R. *Lone Star: A History of Texas and Texans.* 1969; Da Capo Press, 2000.

Fitzhugh, Lester N. "Saluria, Fort Esperanza, and Military Operations on the Texas Coast 1861–1864." *Southwestern Historical Quarterly* 61, no. 1 (July 1957): 66–100.

Foster, William C. *Historic Native Peoples of Texas.* Austin: University of Texas Press, 2008.

Gallaway, B. P, ed. *Texas, the Dark Corner of the Confederacy: Contemporary Accounts of the Lone Star State in the Civil War.* 3rd ed. Lincoln: University of Nebraska Press, 1994.

Graves, Joseph A. *The Alstons and the Allstons of North and South Carolina* (genealogy). Atlanta: Franklin Printing and Publishing Company, 1901.

Guthrie, Keith. *Texas Forgotten Ports.* Austin: Eakin, 1988.

Hawkins, John Davis. "Colonel Philemon Hawkins, Sr." 1829, 1906; repr. in *North Carolina Booklet* (Raleigh: North Carolina Society Daughters of the Revolution, Commercial Printing Company), vol. 19, no. 1–2 (July–October 1919): 92–106.

Holbrook, Abigail Curlee. "A Glimpse of Life on Antebellum Slave Plantations in Texas." *Southwestern Historical Quarterly* 76, no. 4 (April 1973): 361–83.

The Handbook of Texas. 3 vols. Vols. 1, 2 (1952) edited by Walter Prescott Webb, vol. 3 (1976) edited by Eldon Stephen Branda. Austin: Texas State Historical Association. *The Handbook of Texas Online,* http://www.tshaonline.org.

Ingram, Mary McAllister. *Canebrake Settlements: Colonists, Plantations, Churches, 1822–1870, Matagorda County, Texas.* Bay City, Tex.: Lyle Printing, 2006.

Jeter, Lorraine Bruce. *Matagorda, Early History.* Baltimore: Gateway Press, 1974.

Johnson, William R. *A Short History of the Sugar Industry in Texas.* Texas Gulf Coast Historical Association Publications, vol. 5, no. 1. Houston: Texas Gulf Coast Historical Association, April 1961.

Jones, C. Allan. *Texas Roots: Agriculture and Rural Life before the Civil War.* College Station: Texas A&M University Press, 2005.

Kilman, Ed. *Cannibal Coast.* San Antonio: Naylor, 1959.

Lamar, Howard R. *Charlie Siringo's West: An Interpretive Biography.* Albuquerque: University of New Mexico Press, 2005.

La Vere, David. *The Texas Indians.* College Station: Texas A&M University Press, 2004.

Lewis, Frank Hawkins. "The Range Cattle Industry in the Southwest." Senior thesis in economics, Princeton University, 1943.

Lienhard, John H. "Almira Hart Lincoln Phelps." Houston KUHF radio series Engines of Our Ingenuity, episode 2032, College of Engineering, University of Houston.

Marr, John Columbus. *The History of Matagorda County, Texas.* Master's thesis, University of Texas, 1928.

Massey, Sara R., ed. *Black Cowboys of Texas.* College Station: Texas A&M University Press, 2000.

Matagorda County Historical Commission. *Historic Matagorda County.* Edited and compiled by Frances V. Parker and Mary B. Ingram. Houston: D. Armstrong, 1986.
Newcomb, W. W. Jr. *The Indians of Texas* Austin: University of Texas Press, 1961.
Ogilvie, M. B. *Women in Science: Antiquity through the Nineteenth Century.* Cambridge, Mass.: MIT Press, 1986.
Patapsco Female Institute web site, www.ellicottcity.net/tourism/attractions/patapsco.
Ricklis, Robert A. *The Karankawa Indians of Texas: An Ecological Study of Cultural Tradition and Change.* Austin: University of Texas Press, 1996.
Salamon, Sonya. *Prairie Patrimony: Family, Farming, and Community in the Midwest.* Chapel Hill: University of North Carolina Press, 1992.
Schiewitz, E. M. *Buck Schiewetz' Texas: Drawings and Paintings by E. M. Schiewetz.* Introduction by Walter Prescott Webb. Austin: University of Texas Press, 1960.
Schneider, Herbert W., ed. *The Autobiography of Benjamin Franklin.* Indianapolis: Bobbs Merrill, 1952.
Silverthorne, Elizabeth. *Plantation Life in Texas.* College Station: Texas A&M University Press, 1986.
Siringo, Charles A. *A Texas Cowboy, or Fifteen Years on the Hurricane Deck of a Spanish Pony.* 1886; New York: Penguin Classics, 2000.
Stieghorst, Junann. *Bay City and Matagorda County.* Austin: Pemberton Press, 1965.
Sunday Express News (San Antonio). Interview with Arthur Green. February 2, 1986.
The Texas Almanac. Distributed by Texas A&M University Press, www.texasalmanac.com.
Weber, Max. *The Protestant Ethic and the Spirit of Capitalism.* 1905 (in German); English translation by Talcott Parsons, London: Unwin Hyman, 1930.
Wilbarger, John Wesley. *Indian Depredations in Texas.* 1889; reprint, Austin: Pemberton Press, 1967.

Index